CAMBRIDGE LIBRARY COLLECTION

Books of enduring scholarly value

Technology

The focus of this series is engineering, broadly construed. It covers technological innovation from a range of periods and cultures, but centres on the technological achievements of the industrial era in the West, particularly in the nineteenth century, as understood by their contemporaries. Infrastructure is one major focus, covering the building of railways and canals, bridges and tunnels, land drainage, the laying of submarine cables, and the construction of docks and lighthouses. Other key topics include developments in industrial and manufacturing fields such as mining technology, the production of iron and steel, the use of steam power, and chemical processes such as photography and textile dyes.

On the Power of Machines

From the 1770s onwards, John Banks (1740–1805) taught natural philosophy and gave courses of public lectures across the north west of England. Much of his work aimed to show engineers, mechanics and artisans how they could benefit from expanding their practical and theoretical knowledge. In this 1803 publication, Banks ranges across mechanics, hydraulics and the strength of materials. He considers various designs for important industrial machines, such as watermills, pumps and steam engines, offering calculations of their power. Drawing on his own experiments, as well as those of others, he shows readers how to estimate the strength of wooden and iron beams, and how to calculate the airflow from a pair of bellows. Diverse in its topics, the book sheds light on how rational calculation came to be applied to the machinery of the industrial revolution. Banks' *Treatise on Mills* (2nd edition, 1815) is also reissued in this series.

Cambridge University Press has long been a pioneer in the reissuing of out-of-print titles from its own backlist, producing digital reprints of books that are still sought after by scholars and students but could not be reprinted economically using traditional technology. The Cambridge Library Collection extends this activity to a wider range of books which are still of importance to researchers and professionals, either for the source material they contain, or as landmarks in the history of their academic discipline.

Drawing from the world-renowned collections in the Cambridge University Library and other partner libraries, and guided by the advice of experts in each subject area, Cambridge University Press is using state-of-the-art scanning machines in its own Printing House to capture the content of each book selected for inclusion. The files are processed to give a consistently clear, crisp image, and the books finished to the high quality standard for which the Press is recognised around the world. The latest print-on-demand technology ensures that the books will remain available indefinitely, and that orders for single or multiple copies can quickly be supplied.

The Cambridge Library Collection brings back to life books of enduring scholarly value (including out-of-copyright works originally issued by other publishers) across a wide range of disciplines in the humanities and social sciences and in science and technology.

On the Power of Machines

John Banks

CAMBRIDGE
UNIVERSITY PRESS

University Printing House, Cambridge, CB2 8BS, United Kingdom

Cambridge University Press is part of the University of Cambridge.
It furthers the University's mission by disseminating knowledge in the pursuit of education, learning and research at the highest international levels of excellence.

www.cambridge.org
Information on this title: www.cambridge.org/9781108070270

This edition first published 1803
This digitally printed version 2014

ISBN 978-1-108-07027-0 Paperback

Additional resources for this publication at www.cambridge.org/9781108070270

ON THE

POWER OF MACHINES,

ETC.

INCLUDING

DOCTOR BARKER'S MILL, WESTGARTH'S ENGINE, COOPER'S MILL, HORIZONTAL WATER-WHEEL, CENTRIFUGAL PUMP, COMMON PUMP, ETC.,

WITH THE METHOD OF COMPUTING THEIR FORCE.

DESCRIPTION

OF

A SIMPLE INSTRUMENT FOR MEASURING THE VELOCITY OF AIR OUT OF BELLOWS, WHATEVER BE THE PRESSURE.

ALSO OF

A GAUGE FOR MEASURING WITH ACCURACY THE RAREFACTION IN THE CYLINDER OF A STEAM ENGINE

DEMONSTRATION OF A PARALLEL MOTION.

OBSERVATIONS

ON

WHEEL CARRIAGES, ON LATHES, ON THE LEVER, ETC.

CONSTRUCTION OF A CRANK,

WHICH WILL ACT OR BE ACTED UPON NEARLY WITH THE SAME FORCE IN EVERY PART OF A REVOLUTION, EXCEPT TOP AND BOTTOM.

EXPERIMENTS

ON

THE STRENGTH OF OAK, FIR, AND CAST IRON;

WITH MANY OBSERVATIONS RESPECTING THE FORM AND DIMENSIONS OF BEAMS FOR STEAM ENGINES, ETC.

By JOHN BANKS,

LECTURER ON PHILOSOPHY.

Kendal:

PRINTED BY W. PENNINGTON;

AND SOLD BY W. J. AND J. RICHARDSON, LONDON.

1803.

PREFACE.

SOME of the machines introduced into the following work have been much extolled by different writers, who, at the ſame time, have given us no reaſon why they ought to be recommended; as they have not given us any juſt rules whereby to compute their force.

By the following work it appears that *Weſtgarth's Machine*, the *Centrifugal Machine*, and Dr. Barker's Mill, come far ſhort of the powers which ſome have aſcribed to them, as alſo the Old Mill at Nuneaton, ſo much applauded. It will readily be granted that the three firſt are ingenious inventions, and may in certain ſituations be uſeful, as any other machines, eſpecially the *Centrifugal Pump*: the advantage of which con-

ſiſts

ſiſts in its having the water conſtantly riſing, or a continued ſtream aſcending while it is continued in motion, ſo that, different from a piſton pump, the *vis inertia* of the column of water is not to put in motion at every ſtroke.

Dr. Barker's mill cannot be recommended in practice, for, though it is preferable to an underſhot wheel, it cannot well be applied where there is a larger ſtream, as the other generally is, and if the ſtream is ſmall, an overſhot may be applied to much greater advantage. If there is little work to be done, and a ſufficient quantity of water, it, no doubt, may be recommended on account of the eaſy expence at which it is erected, compared with a bucket wheel.

The reader will obſerve that I have made many experiments on this mill, but have ſtill left the theory imperfect. What I have done may poſſibly be of ſome uſe to the next who may turn his attention to it.

The problem concerning the Lathe is accurate, but not by any means what I could wiſh. Some perſon better acquainted with ſcience may perhaps give a much ſhorter and ſimpler ſolution, but it is the beſt I could give.

The

The experiments on the velocity of air, for any thing I know, are new, as alſo the application of the inſtrument for meaſuring the velocity with which it iſſues out of bellows. The application of the ſteam gauge I preſume is new alſo. The experiments on the ſtrength of wood and caſt iron I hope will be found uſeful, eſpecially in their application to beams for different purpoſes; I would gladly have added many more, but circumſtances would not permit. Some gentlemen of fortune and ingenuity may perhaps continue them for their own and the public good.

I beg leave to obſerve to my reader, that if the work was to be reprinted it might be put in a better form, but at preſent I muſt ſubmit it to his candour. The figures in the plates were engraved, but not numbered till the engraver had a printed copy of the work, hence, the numbers in the different plates do not ſucceed in a regular manner, but as they are right numbered they will eaſily be found; only fig. 5, of which there are two, both belonging to Weſtgarth's engine, one in plate 1, the other in plate 3.

ERRATA.

ERRATA.

Page 20, *read* Fig. 5, plate 1 & 3; *p.* 49, *l.* 5, *for qs r. pv*; *p.* 56, *the following table or ſcheme is omitted*

CD.	X.	CE.
$r = .5$	.369	.869
$r = .6$	.312	.912
$r = .7$	.254	.954
$r = .8$	.194	.994
$r = .9$	.133	1.033

p. 63, *l.* 6, *for* CE *r.* CC; *l.* 8, *for* CDE *r.* CDr; *p.* 92, *l.* 12, *for* plate 4 *r.* plate 3; *p.* 104, *l.* 10, *for* axles *r.* axle.

CONTENTS

CONTENTS.

Of

ON

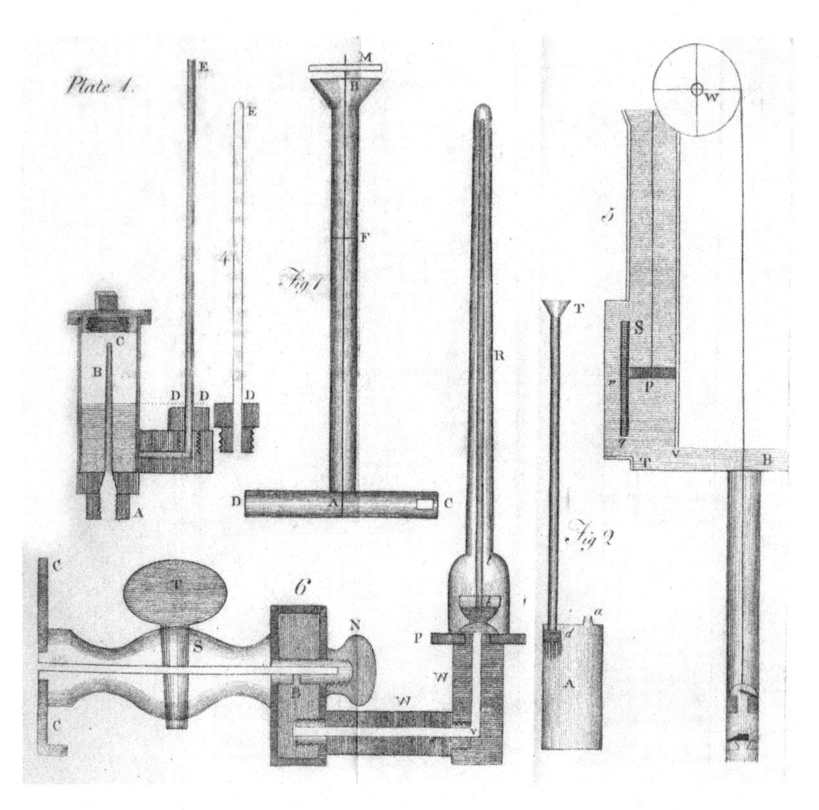

The material originally positioned here is too large for reproduction in this reissue. A PDF can be downloaded from the web address given on page iv of this book, by clicking on 'Resources Available'.

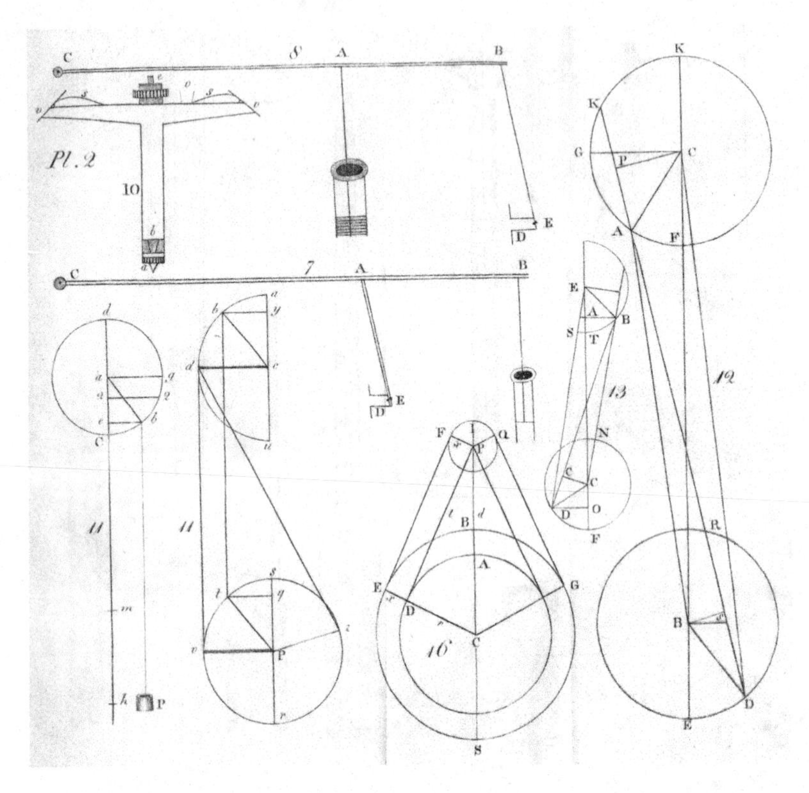

The material originally positioned here is too large for reproduction in this reissue. A PDF can be downloaded from the web address given on page iv of this book, by clicking on 'Resources Available'.

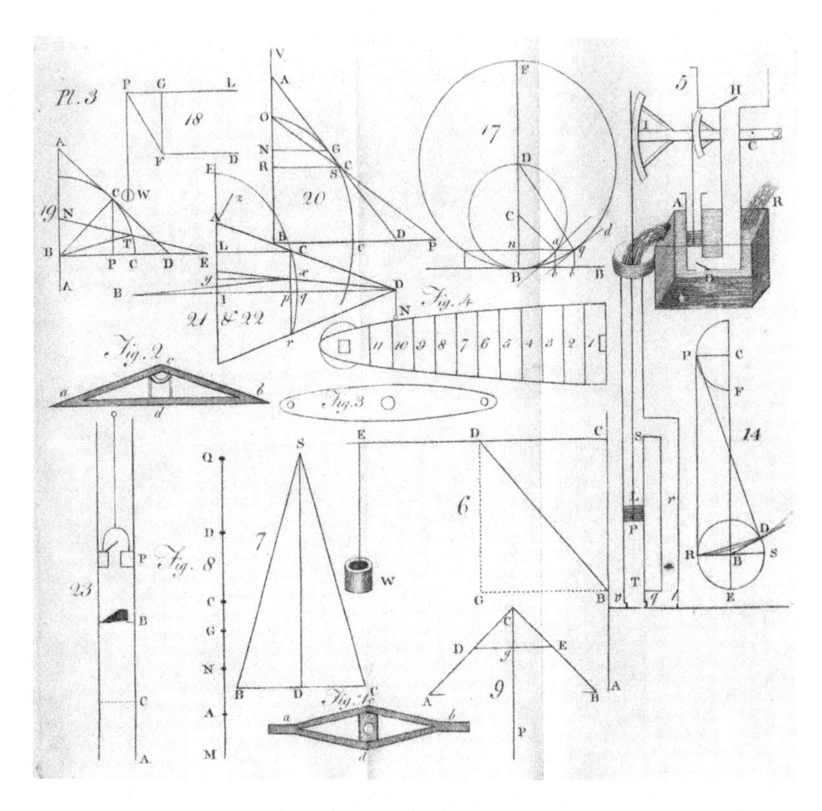

The material originally positioned here is too large for reproduction in this reissue. A PDF can be downloaded from the web address given on page iv of this book, by clicking on 'Resources Available'.

ON THE

POWER AND PROPERTIES

OF

DR. BARKER'S MILL.

THIS mill conſiſts of an upright tube, A B, Fig. 1, round or ſquare, communicating with two arms A D, and A C, which have openings on oppoſite ſides, and which may be more or leſs cloſed by ſlides as at S. That in the arm D is hid on the other ſide. A ſtream of water falls in at the top at B. The whole turns round on the pivot A, and is kept upright by the axle F B, which is fixed into the croſs pieces at F and B, and alſo may ſupport a millſtone, M.

On the cauſe of motion.

When the water has filled the arms and trunk, the preſſure of the water againſt the ſides of the horizontal arms will be directly as the altitude of the water A F, or A B. If the altitude A B is 12 feet, then the preſſure againſt the ſides of the arms will be 5.2 pounds on every ſquare inch.

Hence, if a hole one inch ſquare is cut in each arm, and on oppoſite ſides, and if the ſtream of water is ſufficient to keep the mill filled up to B, then the preſſure againſt thoſe ſides which are not cut, will be 5.2 pounds more than againſt the perforated ſides; hence we ſhall have 10.4 pounds acting on the arms to carry the mill round.

Additional power.

When motion is communicated to the mill by the power above deſcribed, another power begins to act in conjunction therewith, that is, the centrifugal force of the water in the horizontal arms, which is acquired by the whirling motion, and, like an additional head, cauſes the water to preſs with more force againſt the ſides of the arms. This power increaſes as the ſquare of the velocity, and, when that is known, may eaſily be compared with the firſt power, as ſhall be ſhewn hereafter. But, as the velocity increaſes, the reſiſtance increaſes, till theſe powers combined can no longer accelerate the motion, hence it becomes uniform; but at what degree of velocity is not to be aſcertained, but by experiment; becauſe the friction and different ſorts of reſiſtance cannot be accurately computed.

To

To compute the centrifugal force.

Let a = A C, or A D = 3 feet,

T = time of revolving = $\frac{1}{2}$ fecond.

$a\sqrt{\frac{1}{3}} = 1.732$ the centre of gyration of the arm, that is its diftance from A, which, doubled, will be the diameter of a circle, in which the whole weight of both arms may be confidered as acting; therefore if D = 3.464 and $q = 3.1416$, the centrifugal force will be expreffed by $\frac{Dq^2}{ST^2}$; (where S = 16 feet) and $\frac{33.9472}{4} = 8.4868$ the centrifugal force compared with the weight, or quantity, or length of the arm; and, as the diameter of the arm will make no difference in the preffure againft a given fpace, the quantity may be reprefented by a, the lengh of the arm, and $a \times 8.4868 = 25.46$ feet; fo that the preffure is equal to the original head added to this, viz. to $12+25.46$, or to 37.46 feet, or with this additional force, the velocity of the water from the arms is as great, with a head of 12 feet, as it would be with a head of 37 feet, if the mill was at reft.

This is perhaps a greater difference than might take place in practice; but the following experiments, made on a fmaller fcale, will fhew that the power when the mill is moving, is much greater than when it is at reft.

Experiments.

Altitude of the trunk - - - 1 foot $= b$.

Length of one arm - - - - - .56 foot $= a$.

Diameter of circle of gyration $2 \times \cdot 323 = d =$ 646,

Time of a revolution - - - $\frac{3}{8}''$ $= \text{T}$,

When the mill is at reſt, it diſcharges 6.1 pounds per minute. When revolving 160 times per minute, 9.87 pounds per minute.

Then $\frac{\text{D}q^2}{\text{ST}^2}$ or $\frac{1.154\,q^2}{16\,\text{T}^2} = 2.835$, which multiplied by the length of the arm .56 gives 1.585 for the additional force, which added to the other, is equal to a head of 2.585 feet; otherwiſe, the quantity of water diſcharged through given openings, being always as the ſquare root of the depth, we may ſay as 6.1 is to the ſquare root of 1 foot, ſo is 9.87 to the ſquare root of the depth, which would produce that velocity, viz.

$$6.1 : 1 :: 9.87 : 1.618;$$

the ſquare of which is 2.617 feet, for the head: differing only .032 foot from that computed by the laws of circular motion, and which ſmall difference may ariſe from the inaccuracy of the dimenſions, or of obſerving the time, or both.

Experiments

With a mill in which the water was kept at the height of 22 inches, and weights raiſed 6 feet 10 inches, by one gallon of water per minute.

Length

Length of the arms 5½ inches.

No.	Weight raised.	Time in seconds.	Velocity of the wt. in inches.	Effect.	Velo. of the holes in feet.	Head in feet.	Velo. of the water.
1	1½	8½	9.64	14.46	5.759	2.43	8.37
2	2	11	7.45	14.9	4.466	2.19	7.99
3	2½	15	5.46	13.65	3.263	2.01	7.61
4	600 turns without weight in 2′ 12″.						
5	600 turns with a fly 1 pound weight and 15 inches diameter in 3′ 8″.						
	Length of the arms 7½ inches.						
6	2	10.2	8.03	16.06	6.54	2.606	8.694
7	2½	12½	6.56	16.4	5.34	2.351	8.262
8	3	14¾	5.46	16.38	4.52	2.202	7.992
9	600 turns in 2′ 50″						
10	600 turns with the fly, in 3′ 22″						
	The following with arms 19.6 inches. The weights raised 6 feet.						
11	3	18½	3.89	11.67	8.208	3.046	9.396
12	4	20	3.4	13.6	7.592	2.873	9.126
13	6	22	3.27	19.62	6.891	2.688	8.826
14	8	26	2.77	22.16	5.831	2.444	8.424
15	10	28	2.57	25.7	5.42	2.362	8.262
16	12	52½	1.37	16.44	2.89	1.983	7.603
17	8	27	with the fly.				
18	600 turns in 8′ 20″						
19	600 turns with the fly in 9′						

N. B. 17 Turns of the mill raises the weight 6 feet 10 inches.

It

It is not certain that in any of theſe experiments the greateſt effect has been produced, for in the firſt claſs it ought evidently to be ſomething leſs than 2 ounces; with the ſecond arms, it appears to be in raiſing between 2 ounces and 2½ ounces, and with the long arms, 10 ounces ſeem to be too much. But admit the 2d, 7th, and 15th experiments to be maximums, then we may infer that the force increaſes with the ſquare root of the length of the arms, and that the velocity of that part of the arms, where the water is diſcharged, is ⅔ of the velocity with which the water is diſcharged, for in this, different from the under-ſhot, the mill cannot leave the preſſure; and in the under-ſhot, at maximum, the floats are ſtruck with ⅔ the velocity of the ſtream, and the wheel moves with ⅓ thereof. Hence they who have ſuppoſed that the arms in this, as in the under-ſhot, ſhould move with ⅓ the velocity of the water diſcharged by them, ſeem, by theſe experiments to have reaſoned falſly.

In the 4th experiment the centrifugal force added to the height of the mill, gives a head equal to 58 inches.

In the 9th experiment the original head and centrifugal force equal to 63.6 inches.

In the 18th experiment the two combined are equal to 54.5 inches.

Whatever

Whatever length of arms be uſed, when the mill is at a maximum, the head by theſe experiments is raiſed about $\frac{2}{7}$ of the height of the trunk, for inſtance in the above, from 22 to 28.2 inches, as in the 2d, 7th, and 15th experiments, a head of 28.2 inches will produce a velocity of 8.1 feet per ſecond, which multiplied by 60 gives 486 feet per minute, the length of a ſtream, the ſolidity of which in this caſe is equal to 282 cubic inches, the area of its ſection is .048353 inch, this ſection multiplyed by 28.2 inches, and reduced into ounces is the force acting at the ends of the arms to turn the mill, and is equal to .788 ounce.

In the firſt five experiments the diameter of the circle of gyration is .529 foot.

In the next five it is .7216 foot.

From the 10th to the 19th it is 1.885 feet.

To find the centrifugal force of the water in the arms.

Let D = the diemeter of the circle of gyration.
q = 3.1416.
S = 16.
T = time of a revolution.

Then $\frac{Dq_2}{ST^2}$ × the length of one arm, gives the altitude of a column, the preſſure of which is equal to the centrifugal force.

EXAM-

EXAMPLE.

$$\text{D} = .529 \quad \text{Log. } 1.7234557$$
$$q^2 = \quad .9943018$$
$$\text{D}q^2 = \quad .7177575$$
$$\text{S} = 16 \quad 1.2041200$$
$$\text{T}^2 = .25 = \quad \bar{1}.3979400$$
$$\text{ST}^2 = \quad .6020600$$
$$\text{D}q^2 - \text{ST}^2 = \quad .1156975 \quad \bar{1}.6611783$$

The length of one arm is .45833 foot — $\bar{1}.7768758 = .59834$,

which added to the height of the mill gives 2.43167 feet for the head.

ON

ON THE

VELOCITY OF AIR

FROM BELLOWS, &c.

A Tube filled with any kind of fluid, as AIR, WATER, QUICKSILVER, &c. and placed in a vacuum, every kind of fluid would flow out with the ſame velocity; for if quickſilver be heavier than water, and of conſequence produce a greater preſſure, yet the particles to be impelled are heavier alſo, and require force in proportion to their weight to project them with the ſame velocity as water. And if air is lighter than water, the particles to be projected are alſo lighter, hence the velocity by a given head will be the ſame in all.

A tube 16 feet long filled with air, would, like water, flow out with a velocity of 32 feet per ſecond, making no corrections for reſiſtance.

And if we take the gravity of air and water as 840 to 1, or if water is 840 times heavier than air

air, then one foot of water compreſſing air, would produce as great a velocity as 840 feet of air, or as a column of air 840 feet high could do by its gravity.

If we take the whole preſſure of the atmoſphere equal to 33 feet of water, or its height (ſuppoſing it to be equally denſe, which will make no difference in this caſe) equal to 33×840, equal to 27720 feet. Then as the ſquare root of 16 is to 32 feet velocity, ſo is the ſquare root of 27720 to 1332 feet, the velocity with which the air would enter into an indefinite vacuum.

To prove whether air compreſſed by 32 feet of water would be impelled into the atmoſphere with the above velocity, I made, amongſt many more, the following experiments.

Fig. 2.

A, is a caſk of a known capacity, into the top of which is ſcrewed an aperture *a*, of a known area.

The tube T *d*, recurve at *d*, is ſoldered or ſcrewed into the top of the ſaid caſk.

The hole *a* is ſtopped, and water poured into the tube at T, till it is full, at which time a quantity of water will have paſſed out of the tube at *d*, and

and condenſed the air in the caſk till its ſpring is equal to the weight of the water in the tube.

At this time a cock placed over the tube T, ſufficiently large to ſupply water as faſt as it can deſcend into the veſſel A, muſt be opened to keep the tube conſtantly filled; for this purpoſe one perſon muſt attend it, and another muſt open the aperture *a*, which need only have been cloſed by a finger, and he muſt meaſure the ſeconds from the moment that the finger is removed, till the water flies out at the jet. Hence, from knowing the capacity of the veſſel, and the area of the jet, the velocity may be obtained.

If the tube T *d*, ſhould be continued near the bottom of A, while A was filling with water, the length of the compreſſing column would be gradually diminiſhed, and of conſequence the preſſure would be conſtantly changing; hence the open end of the tube is as near the top of the caſk as is conſiſtent with a free paſſage for the water.

Experiment.

The veſſel A contained 15 lb. 6 oz. of water; from which we find its capacity 425.088 cubic inches.

The area of the aperture *a*, through which the air is diſcharged, is .0046 inches.

Ex. 1.

Ex. 1. { The altitude of T above the caſk, 30 in.; Time of expelling the air 33″; by ſeveral trials.

Ex. 2. { The altitude of T 6 feet; Time of expelling the air 21.3″; by ſeveral trials.

In the firſt experiment 425.088, the ſolidity or cubic inches of air in the caſk, being divided by 0046, the area of the hole, gives 92410.4 inches for the length of the ſtream driven out in 33″, by which divide that length, and we have 233.3 feet, the velocity per ſecond.

The ſecond experiment by the ſame proceſs gives 361.6 feet per ſecond; and if we ſhould find this from the firſt, we may ſay, as the ſquare root of $2\frac{1}{2}$, the head, is to 233.3, ſo is the ſquare root of 6 feet, the head, to 361.8 feet.

To compare the velocity, thus found by experiment, with that aſſigned by theory (page 9) we may ſay, as the ſquare root of 6 is to 361.6, the velocity produced by that head, ſo is the ſquare root of 33 to 845.2 feet per ſecond, the velocity produced by that head; or the velocity with which the atmoſphere would begin to enter into a vacuum.

But the velocity thus found by experiment, is, for a head of 33 feet, 487 feet per ſecond leſs than that found by theory, but if we chooſe to compute by theory, and make the ſame corrections as as for flowing or ſpouting water, the reſult will nearly agree with the experiment, viz. if we multiply the compured velocity by .634, the product will be nearly the true velocity; for example, $1332 \times .634 = 844.5$.

It has been propoſed, and attempted, to find the velocity of air by letting a piſton of known weight deſcend in a perpendicular cylinder with an aperture of known area; the objection is, that the friction diminiſhes the weight of the piſton, or ſome air makes its eſcape, which renders the concluſion uncertain.

I have alſo made experiments by ſinking veſſels in water till their tops were even with its ſurface, and then opening the aperture, that the riſing water might expel the air, by which I obtained the ſame velocities as above; but the method of computing here is much more intricate than the other, for which reaſon I ſhall not inſert the experiments.

To find the velocity with which air is driven out of bellows, of any form or ſize, and loaded with unknown weight, and without knowing the area of the pipe or aperture through which it is diſcharged.

Fig. 3.

B is a tube of braſs, or iron, about the ſize of the figure; a bottom may be ſcrewed or ſoldered to it, it, but not with soft solder; A C, a tube made fast in that bottom, and reaching near the top.

The top plate L must be fixed with a screw, or screws and leather.

D E is a pretty strong glass tube, but the bore need not be more than one eighth of an inch, though this is not of consequence; its length may be 3 feet or more. When the instrument is to be used, it is to be filled with water to within one inch of the top, and the end A may be stuck fast into a hole made to receive it in the upper board of common smith's bellows; then, blowing gently, the pressure will act upon the surface of the water contained in the instrument, and cause it to rise in the tube D E, which is open at top; and by the altitude of the water, we know, from the table, the velocity of the air leaving the bellows.

When the air requires to be much compressed, as in iron furnaces, the water would flow out at the top of the tube, except it were five feet or more in length, in which case quicksilver may be used instead of water, and the tube may be shorter, as for instance, one foot would be long enough for the strongest blast, or greatest compression of air in any kind of bellows. The condensation is seldom greater than that which would be produced by 4 inches.

Table

TABLE

Of the Velocity communicated to Air by Water, &c.

Altitude of Mercury.	Altitude of Water.	Velocity of Air in Feet per Second.	Altitude of Mercury.	Altitude of Water.	Velocity of Air in Feet per Socond.
Inches.	Inches.		Inches.	Feet.	
.077	1	42	8.3	9	442
.15	2	57	10.	12	510
.23	3	73	13.	14	550
.30	4	85	14.	16	589
.38	5	95			
.46	6	104	Feet.		
.53	7	112	1.3	17	608
.61	8	123	1.38	18	622
.69	9	128	1.46	19	638
.77	10	134	1.53	20	658.5
.84	11	141	1 61	21	672
.92	12	147¼	1.7	22	690.6
			1.77	23	704
1.15	15	164	1.84	24	719
1.38	18	180	1.92	25	736
1.61	21	194	2.	26	748
			2.07	27	763
	Feet.		2.15	28	778.9
1.84	2	208	2.23	29	789
2.8	3	251.8	2.3	30	803
3 7	4	294	2.38	31	818
4.15	4.5	312	2.46	32	832
4.61	5	329	2.53	33	845
5.53	6	360			

By the table and inſtrument it is eaſy to compare the velocities communicated to air by different

rent bellows, whether we uſe mercury or water. In ſmall velocities, water will be moſt convenient, as the ſcale will be larger; but when the preſſure is equal to 3 or 4 feet of water, on account of the length of tube, mercury will be more convenient. The inſtrument or gage may be connected with any part of the bellows or tube, where the air is condenſed; a hole may be bored through the iron near the tewel, or noſle of the bellows, as a proper place to try the condenſation; and ſuppoſe quickſilver is raiſed 5.5 inches, then oppoſite is 360 feet for the velocity of the air per ſecond; or 245 miles in one hour.

Fig. 4.

Inſtead of a long tube, for a ſtrong blaſt, the end E may be ſealed, in which caſe a tube of 12 inches long may be ſufficiently accurate, and may have a ſcale adapted to it. Suppoſe it filled with water to D D, and the top L ſcrewed tight, then whatever quantity of air may be forced through A C into the gage B, it will preſs upon the water, and cauſe it to riſe in the tube D E till the confined air is of the ſame denſity as that in B; that is, if the denſity in B is doubled, then that in D E will alſo be doubled, or the water will have filled half the tube. In ſhort, the denſity in B, and the upper part of the tube D E, will always be equal. And it will be reciprocally, as the length of the tube

tube is to the preſſure of the atmoſphere, ſo is the length of the column of compreſſed air to the force which compreſſed it. Suppoſe the water has riſen one inch in the tube, then as $12:1::11:1\frac{1}{11}$, viz. the air is compreſſed by one atmoſphere and $\frac{1}{11}$ of the atmoſphere. Hence, to make a ſcale for a tube of any given length whatever. If we divide the whole length of the tube, by the length of that part in which the air is compreſſed, the quotient will be the compreſſing force in terms of the atmoſphere, viz. it may be an atmoſphere and a tenth, an eight, a quarter, a half, or 2 or 3 atmoſpheres, and parts, &c.

Otherwiſe, If we take the preſſure of the air equal to 32 feet of water, then multiply the length of the tube by 32, and divide the product by the length of the part containing compreſſed air, and the quotient will be a number from which, if 32 be ſubtracted, the remainder will be the head of water which would produce that compreſſion. Suppoſe, as before, that the water has riſen one inch in the tube, then as $12:32::11:34\frac{10}{11}$; from which take 32 and we have $2\frac{10}{11}$ feet of water to produce that compreſſure, equal to $\frac{1}{11}$ of the whole atmoſphere.

If $t =$ the length of the tube in inches.

$h =$ the height of the column of water equal to the preſſure of the atmoſphere, in feet.

n

D

$n =$ the length of compreſſed air in the tube, in inches.

$b =$ the height of water which produces that compreſſion, in feet.

Then as $t : h :: n : \frac{th}{n} = b + h$;

Hence $b = \frac{th}{n} - h$ Theo. 1, if the atmoſphere is repreſented by 1.

$$n = \frac{th}{h + b} \quad \text{Theo. 2,}$$

Then $b = \frac{t}{n} - 1$ Theo. 3.

If the tube in which the water riſes is of conſiderable length, and the preſſure great enough, to raiſe the water ſome feet, the altitude of the water ought to be taken from the height of the atmoſphere; or, in computing, h ought to be encreaſed as much as the height of the water in the tube.

But if the tube is only 12 or 18 inches long, any altitude to which water will be raiſed by any common blaſt, whether for forges or furnaces, is not worth notice, eſpecially as the weight of the atmoſphere is continually varying. If mercury ſhould be uſed, it would produce a greater error, and, in ſome caſes, muſt be taken into the account.

Ex.

Ex. 1. If, in a tube 12 inches long, the air is compreſſed into 9 inches, by quickſilver, what is the compreſſing force, the height of the barometer being 30 inches?

Theo. 3. $\frac{t}{n} = b$ viz. $\frac{12}{9} = 1\frac{1}{3}$, to which add $\frac{3}{30}$, or $\frac{1}{10}$, for the altitude in the tube, and we have $1\frac{1}{3} + \frac{1}{10} = 1.\frac{13}{30}$ atmoſpheres for the whole preſſure, or near one atmoſphere and a half.

Ex. 2. Suppoſe when the barometer ſtands at 29 inches, that quickſilver is raiſed 7 inches in a tube 20 inches long; required the preſſure?

In this caſe we have 13 inches of air, and $\frac{20}{13}$ gives $1\frac{7}{13}$, to which add $\frac{7}{29}$ and we have $1\frac{294}{377} =$ 1.78, or almoſt two atmoſpheres; or, if multiplied by 29, the height of the barometer, gives 51.6 inches, viz. the preſſure is equal to a column of 51.6 inches of mercury.

If an upright tube is partly filled with air, and part with mercury, then under the common preſſure of the atmoſphere, the air in the tube will be expanded.

A
DESCRIPTION
OF
WESTGARTH'S ENGINE.

Fig. 5.

THIS figure deviates in form, but not in principle, from Mr. Weſtgarth's model, depoſited in the room of the ſociety for the encouragement of arts, &c. London. I believe but few of theſe engines have been erected; perhaps for want of ſituations, as the engine, where known, is not in want of reputation. I have been told that Mr. Smeaton obſerved that it was the beſt invention we have had ſince that of the ſteam engine. I ſhall firſt explain the engine, and then enquire into its powers

S T is a bored cylinder, with two valves at the bottom *q* and *v*.

P a ſolid piſton, ſuſpended from the end of the beam I C.

r a pipe which connects the water above the piſton with that below, when the valve *q* is open. This cylinder is ſupplied with water from the ciſtern C D, which is itſelf ſupplied from the ſtream R.

A is a forcing pump, placed in the ciſtern, with its piſton rod fixed to the middle of the beam I C, and by which water is forced into the ciſtern H.

Operation.

When the piſton P is up at L, and the valve *v* open and *q* ſhut, the water upon the piſton will carry it down, and will alſo force down the piſton of the forcing pump A, the valve *q* will open and *v* ſhut, a communication will be opened between the water above and below the piſton, ſo that there is nothing to keep the piſton down but its weight, which will, in part, be overcome by the water in the ciſtern C D entering in at the valve, and preſſing againſt the bottom of the piſton in the forcing pump, and partly by a weight O fixed on the other ſide of the centre C. When P has again riſen to L, the valve *q* is ſhut, and *v* opened; by which the water in the lower part of the cylinder has liberty to run out, and therefore cannot ſupport the piſton P, which again deſcends by the preſſure of the water upon it, &c.

Computation.

Computation.

Let the length of the cylinder S T be 40 feet, its area 100 inches, and let the height of H be 40 feet above the piston at A, and the area of A = 124 inches. The weight on the piston P will be $100 \times 40 \times 12 = 48000$ cubic inches of water.

The pressure against the piston at A will be = $124 \times 40 \times 12 = 59520$ cubic inches; as this piston is acted upon by the middle of the beam, had its pressure been 96000 it would only have been a balance for the weight on P; hence P has an advantage of 36480 to overcome the friction and produce a proper velocity. (See my Treatise on Mills, part second.)

Let the descent of P be 6 feet, or 72 inches, which multiplied by its area, 100, gives 7200 cubic inches of water, which is lost at every stroke.

If the descent of P is 6 feet, the descent of A will be 3 feet, or 36 inches, which multiplied by 124, its area, gives 4464 for the cubic inches raised at every stroke. And, as 7200 is to 4464, so is 100 to 62, which the above reference proves to be the greatest effect that can be produced.

Now if 7200 cubic inches, pounds, &c. were placed in a vessel suspended from a rope, chain, &c. passing over a wheel, and if a vessel at the other end of the rope contains 4464, this will certainly

tainly be raifed with as much or more eafe than in an engine, as by this procefs we get clear of the friction of the pumps. *Where then is the advantage?*

In this laft method the buckets would have to fall and rife 40 feet, which would require more time than the other to produce the fame effect, where the length of the ftroke is fuppofed to be 6 feet.

But in refpect of the power there is no advantage: for to raife the fame quantity there is the fame lofs of water: therefore, it would anfwer the fame purpofe to make the buckets larger, which would be done at a much lefs expence.

The great utility of this machine, and to which the inventer, no doubt, had a view, is to raife water out of mines, as thefe, efpecially lead mines, are often in the fides of mountains, it fometimes happens that the water is difcharged from them by levels driven on purpofe, and from thefe levels other fhafts are funk, out of which the water muft be raifed by pumps, or other contrivances.

Suppofe a ftream of water can be brought to the top of the pit, and run into the cylinder S T, which may defcend to the level T B, and if a pump P A is put down the lower pit, it may be wrought by the water in S T, by means of a wheel

wheel and chains, and the water which is employed as the power to work the pump, may run off with the raifed water along T B. This engine does not require much room, and is well fuited for fuch a purpofe, yet all that can be faid for it, is utility; no advantage in point of power belongs to it.

OBSERVATIONS

OBSERVATIONS

ON THE

LEVER.

Fig. 7. 8.

C B, C B are two levers, each 11 feet long $= l$; each is acted upon by a crank D E, the force of which at E is equal to 100 $=$ F.

The ſpear A E, Fig. 7, is fixed to the beam at the diſtance of 7 feet from the centre C, and the pump rod at the extremity, viz. at 11 feet from C.

In Fig. 8, the crank acts at the end of the lever, and the pump-rod is fixed at 7 feet from C. Query, is there any difference in the effects produced by theſe different applications?

Computation.

In the firſt caſe $\frac{AC \times F}{BC} = P$ the power exerted on the end of the lever at B.

And as A C is to 2 D E (the velocity of A) ſo is

CB to $\frac{2DE \times CB}{AC} = V =$ the velocity of B.

And $PV = \frac{AC \times F}{BC} \times \frac{2DE \times CB}{AC}$ the power multiplied by the velocity gives the effect. Let $DE = 1$, then $P = \frac{AC \times F}{BC} = \frac{700}{11} = 63.636$,

and $V = \frac{2DE \times CB}{AC} = \frac{22}{7} = 3.142857$,

alſo $PV = 63.63 \times 3.1428 = 200$, the effect, in the firſt caſe.

Second Caſe.

$\frac{CB \times F}{AC} = P = 157.142857$,

and as $CB : 2DE :: CA : \frac{2DE \times CA}{CB} = 1.2727 = V$

alſo $157.142 \times 1.272 = 200$, the effect, the ſame as in the firſt caſe

Obſervations.

In the firſt caſe, the force exerted by the crank upon the pump rod is - - - - - 63.63,

In the ſecond caſe, it is - - - - 157.142,

The area of a pump raiſing water to a given height may, and ought to be, as the power employed to work it. Hence, in this caſe, the areas may be as the two powers, or the diameters as the ſquare roots thereof. For the areas are as the ſquares

ſquares of the diameters. Or, we may aſſume one diameter at pleaſure, and find the other. Let the diameter of the firſt be 4 inches, to find that of the ſecond, we ſay, as

$63.63 : \overline{4}|^2 :: 157.142 : \overline{6.28}|^2$, the diameter of the ſecond pump.

And $\overline{4}|^2 \times .7854 = 12.56$, the area of the firſt pump.

$\overline{6.28}|^2 \times .7854 = 31.04$, the area of the ſecond pump.

In the firſt pump the velocity of the piſton or length of the ſtroke $= P = 3.1428$.

In the ſecond $P = 1.2727$.

The length of the ſtroke multiplied by the area, gives the quantity of water raiſed at one ſtroke.

And in the firſt, $3.1428 \times 12.56 = 39.43$.

In the ſecond, $1.2727 \times 31.04 = 39.43$.

Let the pump rod in the ſecond fig. be $5\frac{1}{2}$ feet from C, every thing elſe the ſame.

Then $\frac{CB \times F}{AC} = P = \frac{1100}{5.5} = 200$.

And $\frac{2DE \times CA}{CB} = V = \frac{11}{11} = 1$, whence

PV or $200 \times 1 = 200$ the effect, the ſame as the two others.

 To

To find the diameter of the pump.

As $63.63 : \overline{4}|^{2} :: 200 : \overline{7.091}|^{2}$ the diameter.

The length of the ſtroke is 1, which multiplied by 39.43, the area, is exactly equal to the above caſes.

To gentlemen who are acquainted with the properties of the lever, &c. the above obſervations may ſeem unneceſſary. But as there are many artiſts who can erect a pump, a wheel, &c. and yet think very differently reſpecting the above ſimple application of power, I thought that a demonſtration, as ſimple and ſhort as I could draw up, might be of uſe.

ON

COOPER'S MILL.

Fig. 9.

LET the depth of the reſervoir DE be 20 inches.

Water per minute - - - 160 oz.

The area of the opening at D = 1 ſquare inch.

Then will the preſſure upon
the float D - - - = 11.57 oz.

And the quantity of water which may run off without leſſening the head, per ſecond = 2 ⅔ oz

2 ⅔ oz. water is equal to 4.6 cubic inches, and as the aperture is one inch ſquare, the velocity muſt be 4.6 inches per ſecond, to run off as faſt as the reſervoir is ſupplied; of conſequence the velocity of the chain and, of the circumference of the wheel, round which it paſſes, will be the ſame.

In

In ſymbols.

Let d = depth in inches;
a = area of the opening at D;
q = quantity per ſecond in cubic inches;
w = the weight of one in. in oz. = .5787.

Then adw = the preſſure upon the float.

And $\frac{q}{a}$ = the velocity per ſecond, in inches.

This mill compared with an overſhot.

Firſt, let the overſhot be 20 inches diameter, viz. as high as the fall will admit.

Its circumference will be 62.857 inches, and if it move as faſt as the chain in the other, viz. 4.6 inches per ſecond, it will make half a turn in 6.83 ſeconds. And if 60 ſeconds give 160 oz. water, 6.83 ſeconds will give 18.213 oz., with which the wheel is loaded at once, which multiplied .636, to reduce it to the circumference, gives 11.583 oz. the ſame that preſſed upon the chain; ſo that if this chain paſſed round a wheel of 20 inches diameter, the velocities and the powers would be the ſame.

Let the chain paſs over a wheel of 10 inches diameter, and as the velocity of the chain is fixed at 4.6 inches per ſecond, when we are making moſt

moſt of the water, the wheel muſt in this caſe turn round in half the time which it did in the laſt; and has the ſame power acting upon its ſurface, but the radius of this wheel is half that of the former, hence, the power is neither more nor leſs.

This mill, and the overſhot, with reſpect of power are equal; one has no advantage over the other. But with reſpect to ſituation, circumſtances, &c. the overſhot may have advantages which this has not; or this may, in ſome caſes, be preferred to an overſhot, which muſt be left to the judgment of the engineer.

The quantity of water and head being the ſame, if the aperture at D is varied, the following table ſhews the preſſure and velocity of the chain, alſo the weight of water that would be upon a 20 inch wheel, moving with the ſame velocity as the chain, and alſo that weight reduced to the end of the horizontal arm.

The ſecond column in Mr. Cooper's mill is the preſſure upon the chain, and the laſt the force acting on the circumference of the overſhot, which are equal, and which ſhews that the power of one is as great as the other.

Area

Area of D in	Preſſure on D oz.	Velocity of chain.	Weight of, on a 20 in. wheel.	Weight reduced to the end.
.25	2.87	18.4	4.54	2.87
.5	5.75	9.2	8.69	5.75
1	11.57	4.6	17.38	11.57
2	23.14	2.3	34.76	23.14
3	34.5	1 53	52.15	34.50
4	46.	1.15	69.53	46.00

Given, the diameter of the wheel A, and the velocity with which we wiſh it to move, alſo the quantity of water per ſecond, to find the area of the aperture at D.

Let $r =$ the radius of the wheel in feet $= 4$
$v =$ its velocity in feet per ſecond $= 2$
$q =$ cubic feet of water per ſecond $= 10$

Then will $\frac{q}{v} = \frac{10}{2} = 5 =$ the area of D; if we wiſh the floats to be ſquare, extract the ſquare root of 5 for a ſide; if not, we may take any length at pleaſure for one ſide, by which divide 5, and we get the other ſide; ſuppoſe we make one ſide $2\frac{1}{2}$ feet, then the other will be 2, or if one is 3 feet, the other is 1.66.

If $d =$ the depth in feet $= 12$.
$a =$ the area of the aperture in feet $= 5$.
$w = 62\frac{1}{2}$ pounds, the weight of a cubic foot of water.

Then

Then will $adw =$ the preffure on the float at D $= 3750$ pounds. This method may be employed to find the force with which a given ftream would act upon an overfhot wheel, fuppofing the wheel to be made as high as the fall; as for example,

The fall is 18 feet;

The quantity per fecond, 7 cubic feet;

The wheel to move 3 feet per fecond;

Then $\frac{q}{v} = \frac{7}{3} = 2.333 = a$,

and $adw = 2.3 \times 18 \times 62.5 = 2625$ pounds;

or $\frac{qdw}{v} =$ the preffure in pounds.

This is an accurate and expeditious method of computing the force of a ftream of water.

In words.

Multiply the number of cubic feet per fecond, by the depth of the fall, or diameter of the wheel in feet, and multiply that product by 62.5, the weight of a cubic foot of water, and this laft product divided by the intended velocity of the wheel per fecond, gives the force of the water, reduced to the end of the horizontal arm, in pounds avoirdupoife.

A MILL

FORMERLY AT NUNEATON.

THIS mill, the celebrated Benjamin Martin ſays, was reckoned the beſt in England: let us examine it.

The diameter of the water wheel is	16	feet.
Its circumference - - - - - -	50.3	feet.
Head above the wheel - - -	7.5	feet.
Sluice 12 inches wide and 3 inches high, its area - - - - - - -	.25	feet.
The head and fall - - - -	23.5	feet.
The water wheel revolves in -	6.9	ſecon.
Velocity of the circumference per ſecond = - - - - - - -	7 29	feet.
The millſtone 6 feet dia. revolves in	1.29	ſecon.
Force at the ſtones periphery -	2857	pounds
Water per ſec., according to Martin	5.5	feet.
But the true quantity is - - -	3.685	feet.
Velocity of the water per ſecond	14.742	feet.

Weight

Weight, or force of water, on the wheel - - - - - - - - - -	450	pounds.
Impulfe - - - - - - - - -	50	pounds.
Whole force	500	

Cog wheel, 7 feet diameter, contains 48 cogs.

Trundle, 1 $\frac{1}{2}$ feet diameter, has - - 9 rounds $\frac{48}{9} = 5.333$ turns of the millftone, for one of the water wheel.

Obfervations.

According to the above, we have a corn mill erected on a ftream, which has 24 feet of fall, and affords 3.685 cubic feet of water per fecond, which turns a 6 feet millftone 46 $\frac{1}{3}$ times per minute, and this is recommended, by Mr. Benjamin Martin and Mr. James Fergufon as the beft in England; and further, the force of the water at the furface or circumference of the ftone, is 285.7 pounds.

Query. Could a better mill be erected in the given fituation?

Solution $d = 24$ feet
$q = 3.685$ feet
$w = 62\frac{1}{2}$ pound
$v = 3$ feet.

Then $\frac{qdw}{v} = 1842$ lb. the force of the water on the wheel.

Let

Let us ſuppoſe the ſame power acting on the circumference of the ſtone, and enquire how many revolutions the ſtone may make in one minute.

Let b = the radius of the trundle = .8 feet,
h = the radius of the millſtone = 3 feet,
r = the radius of the water wheel = 12 feet,
x = the radius of the cog wheel ſought,
p = the power. - - - - = 1842,
n = the force at the ſurface of the ſtone - - - - = 285.7.

Then $\frac{prb}{hn} = x = \frac{1842 \times 12 \times .8}{3 \times 285.7} = 20.631$ feet, the radius of the cog wheel.

For the radius of the water wheel multiplied by its load, and that product divided by the radius of the cog wheel, gives the force with which the cogs act upon the trundle; that 12 × 1842, and divided by 20.631 gives 1071.4, the force of the water againſt the rounds; which force, multiplied by the radius of the trundle, and divided by the radius of the millſtone, gives 285.7.

The water wheel revolves in 25.14 ſeconds, and the millſtone makes 25.54 turns for one of the water wheel. And as 25.14 ſeconds is to 25.54 turns, ſo is 60 ſeconds to 61 turns. Or, we obtain by this alteration, an addition of 870 turns in one hour, and the quantity of corn ground being nearly

nearly as the ſquare of the velocity, while the mill before ground 11 buſhels, it will now grind 18, and much better.

In this ſolution I have ſuppoſed the ſame power at the circumference of the ſtone, to make the compariſon more ſtriking; but the ſtone would be turned much faſter round with a leſs power, if that power itſelf could get forward; but in the old conſtruction it could not, for the wheel moved nearly as faſt as the water could fall, and in this conſtruction the wheel would move with a greater velocity than is aſſigned to it.

ON

ON THE

HORIZONTAL

WATER-WHEEL.

THESE mills are common in the Isle of Man, and, I am informed, in some parts of the continent.

The wheel is horizontal, and is composed of slips of wood or boards placed in an oblique manner, something like the fly of a smokejack, and are surrounded by a wooden hoop, in order to strengthen them; in all that I have seen the whole wheel does not appear to be worth five pounds; I think it might be made for that sum, or less. On its upright axle, in an upper room, is fixed the millstone, and the water, descending down a sloping trough, strikes the oblique arms, and gives motion to the stone.

Computation.

Let us suppose, as in the last mill, that the fall is 24 feet, and the quantity of water 3.685 feet per second,

ſecond, and that it flows from the ſurface of a perpendicular penſtock, and acts upon the wheel by impulſe only. The ſquare root of 24, = 4.9, which multiplied by 5.4 gives 26.46 feet per ſecond, for the velocity. Then 3.685, the quantity, divided by 26.46, the velocity, gives the area of the ſection, or of the hole through which it flows, equal to 20 ſquare inches, or .1392 of a foot; now 24 × 1392 × 62.5 = 208.8 pounds, the weight of the column, and as 5 : 8 :: 208.8 : 334 pounds, the impulſe of the ſtream upon an object at reſt.

The velocity of the wheel ought to be 8.82 feet per ſecond, one third of the velocity of the ſtream, which, taken from the velocity of the ſtream, leaves 17.64 for the velocity with which the ſtream ſtrikes the wheel, and the force being as the ſquare of the velocity, it will be as $\overline{24}|^2 : 334 :: \overline{17.64}|^2$: 180.43 lb. the force with which the ſtream ſtrikes the wheel in motion. And the force at the ſurface of the ſtone, 285.7, multiplied by 3, the radius of the ſtone, gives 857.1, which, divided by the preſent force, 180.4, gives 4.75 feet, for the radius of the water wheel; hence, its circumference will be equal to 29.8 feet, and the time of a revolution 3.4 ſeconds nearly, or $17\frac{1}{2}$ per minute.

It is evident that in this conſtruction, the water is

is applied to great diſadvantage, and in a country where there is not much corn to grind, and where they have good ſtreams of water, they anſwer the intended purpoſe, and move much faſter than repreſented above.

The cauſe of their greater velocity is, the ſtones are leſs, than in the computation, which is made for the ſake of compariſon, the quantity of water greater, and the water wheel leſs.

CENTRIFUGAL

CENTRIFUGAL PUMP.

Fig. 10.

THE centrifugal machine, or Erſkine's centrifugal pump, as in the figure, conſiſts of an upright tube, and two arms; it has a conical valve *b* at the bottom, and one at the end of each arm, opening outwards, and preſſed againſt the ends of the arms by the ſprings *ss* by a wheel and nut it is turned round on the pivots *a e*.

But, in the firſt place, it is filled with water through the opening or cock at *o*, which is afterwards cloſed, and the ſprings *ss* are juſt ſtrong enough to prevent the valves from being opened by the weight of the water, when the pump is at reſt. The upper ſide of the arms muſt be as high at the ends as the centre, or higher, otherwiſe the air will not be eaſily expelled, but will be apt to lodge under the pivot *e*; the pump is now to be whirled round, and the water flows out at the valves

valves *v v*, but the velocity of the water will vary with the velocity of the pump, with the length of the arms, and with the height to which it is raiſed

Computation.

Let a = the length of one arm in feet.
b = the height to which the water is raiſed in feet.
$y = a\sqrt{\frac{1}{3}}$ = the center of gyration.
$q = 3.1416$.
t = time of a revolution, in ſeconds.
d = 16 feet, the ſpace through which a body falls in one ſecond.
m = 32 feet, the altitude of a column of water, which will balance the weight of the air.

Then will $\frac{2yq^2}{dt^2}$ = the centrifugal force of the water compared with its weight. This force, multiplied by a, the length of the arm, gives $\frac{2ayq^2}{dt^2}$, the length of a column, the preſſure of which is equal to the centrifugal force.

If the altitude b is taken from the depth, the preſſure of which is a balance for the centrifugal force, the ſquare root of the remainder, multiplied by 5.4, gives the velocity of the water flowing from the arms. viz. $= 5.4\sqrt{\frac{2ayq^2}{dt^2} - b}$.

To

To find the time of a revolution when the centrifugal force is equal to the weight, put $\frac{2ayq^2}{dt^2} = b$; from which we shall find $t = q\sqrt{\frac{2ay}{db}}$; or, if $n = \sqrt{\frac{1}{3}} = .5773$, then $t = \frac{aq}{4}\sqrt{\frac{2n}{b}} = .8435a\sqrt{\frac{1}{b}} = \frac{.8435a}{\sqrt{b}}$

Example I.

Let $a =$ 3 feet.
$b =$ 10 feet.
$t =$ 1 second.

Then $\frac{2yq^2}{dt^2} = \frac{3.4638 \times 9.86}{16} = 2.134566$;

the weight of the water may be represented, or expressed, by the length of the arm, therefore $2.134566 \times 3 = 6.403698$ feet, the length of a column which would balance the centrifugal force; but as this is less than the altitude of the pump, it is evident that no water can be raised. For a, and b, being given as above, when the centrifugal force and water in the pump balance each other, $t = \frac{8435a}{\sqrt{b}} = \frac{2.5305}{3.16} = .8002$ seconds.

Example II.

Let $a =$ 3.5 feet.
$b =$ 10 feet.
$t = \frac{1}{2}$ second

Then $\frac{2ayq^2}{dt^2} = 39.86372$ feet,

feet, the length of a column which is equal to the centrifugal force. From which if we take the altitude of the pump, $e\,o$, we have left 29.86372 feet, which, by its preſſure, would produce a velocity equal to that with which the water is diſcharged from the arms, viz. $5.4\sqrt{\frac{2ayq^2}{dt^2} - b} =$ 29.484 feet, the velocity of the water per ſecond.

Let the apertures through which the water flows be two inches each; then ſhall we have $29.484 \times 2 = 58.968$ feet, diſcharged in one ſecond from the two arms; now, if this is multiplied by 12, to reduce it to inches, and then by 2, the area, it will give the cubic inches of water per ſecond = 1415.232 = 5.018 gallon.

If $\left.\begin{array}{r} 5.4 = n \\ \sqrt{\tfrac{1}{r}} = r \end{array}\right\}$ The reſt being given,

$$v = n\sqrt{\frac{2aq^2r^2}{dt^2} - b} = n\sqrt{\frac{a^2 \times 712214}{t^2} - b},$$

$$a = \frac{t}{qn}\sqrt{\frac{dv^2 + dn^2b}{2r}} = .2357 \times t\sqrt{\frac{v^2 + n^2b}{2r}} =$$

$$.22 \times t\sqrt{v^2 + n^2b},$$

$$t = aqn\sqrt{\frac{2r}{dv^2 + dn^2b}} = 4.24186 \times a\sqrt{\frac{2r}{v^2 + n^2b}}$$

Example III.

Given, the altitude to which the water is to be raiſed, 10 feet; the velocity with which we wiſh to to difcharge it, 12 feet per fecond; the time of a revolution, one fecond; required the length of the arms.

Here we have $a = .22 \times t\sqrt{v^2 + n^2 b} = .22 \sqrt{144 + 291.6} = .22 \times 20.8 = 4.577$ feet, for the length of one arm.

Example IV.

Given, the altitude of the pump, 10 feet; the velocity of the water, 12 feet per fecond; the length of the arms, 4.577 each; required the time of a revolution.

$t = 4.555 \times a\sqrt{\frac{1}{v^2 + bn^2}} = 4.555 \times 4.577 \times \sqrt{\frac{1}{435.6}} = 20.848 \times .0479 = .9986$ feconds, anfwer the time of one revolution.

Example V.

Given $\left.\begin{array}{l} a = 4.577 \\ b = 10 \\ t = 1 \end{array}\right\}$ to find v; we have

$n\sqrt{\frac{a^2 \times .712214}{t2} - b} = 5.4\sqrt{\frac{20946 \times 7122}{1} - 10} = 5.4\sqrt{4.9175} = 11.934$ feet, the velocity per fecond.

ON

ON

THE CRANK.

Fig. 11.

SOMETIMES the crank is employed to work pumps, ſaws, &c.; and ſometimes it is acted upon, in order to produce circular motion. Let us firſt conſider it as acting upon a pump rod, &c. Then the crank *a b*, while it moves from *c* to *d*, forces the piſton *p* through *h m*, a ſpace equal to twice its own length, and although the piſton affords the ſame reſiſtance during the ſtroke, yet the power employed to turn the crank, will be very differently reſiſted; for when the end *b* is at *c* or *d* there is no reſiſtance; when at *g* the moſt that it can afford; from *c* to *g*, and from *g* to *d* it is conſtantly varying, being as the ſigns *b e*, 2.2 *g a*, or the whole reſiſtance as the area of the ſemicircle. But if the whole reſiſtance is taken to be equal to that which takes place when the crank

is

is in the pofition *a g*, then would the whole refiftance be expreffed by *c d* × *a g*, which would be falfe.

If a wheel of the fame radius as the crank is furnifhed with teeth to work in a rack, then in half a revolution it will move the pifton through a fpace which is equal to half the circumference of the wheel: for example, let the radius of the wheel, and length of the crank, be half a foot, then will the crank move the pifton 1 foot, but the wheel through half of 3.14159, or 1.57079 feet; hence, the refiftance againft a crank is to the refiftance againft a wheel of the fame fize, as 1 to 1.57079. Or, if we alter the fize of the wheel fo that it may have the fame refiftance as the crank, it is evident that it muft be 2 feet in circumference; for it will then, in half a revolution, move the pifton one foot.

Then 3.14159 : 1 :: 2 : .63662 feet, the diameter of the wheel.

The refiftance againft the wheel is uniform, and in the above cafe is never more or lefs than .6366; whereas a crank, to do the fame work, fometimes has a refiftance equal to 10000, and at other times much lefs.

We may obferve, in computing the force which a crank exerts againft any refiftance when oppofed in the direction of a tangent, it has more force in every

every other poſition, or part of the revolution, as the lever is conſtantly varying in its length: for inſtance, if the force of the crank is 100 pounds, and the reſiſtance 90 pounds, this reſiſtance only takes place when the crank and rod are at right angles with each other, and, in other poſitions, the crank will not feel more than 60, 40, or 20 pounds, &c. to reſiſt it.

If the crank is acted upon by a long bar or rod, which moves nearly parallel to itſelf, then will the acting power meet with the leaſt reſiſtance when it is at right angles with the crank, and with more in every other poſition; at the top and bottom, the crank could not be puſhed forward, except it was furniſhed with a wheel or fly to keep it moving, when no power can be exerted upon it, in order to move it round.

A method by which a power may be applied to act nearly with equal force on a crank, or the crank with equal force againſt the reſiſtance.

Demonſtration.

Let *c d* be an arm fixed upon an axle, moving forward and backward, by a ſteam engine, or any other power, and connected by the bar *dv*, with the crank *p v*; now if *c d* and *p v* be equal in length, then when *d* is at *a*, *v* will be at *s*, and when

when *d c* has deſcribed the angle *a c b*, *p r* will have deſcribed the equal angle *s p t*, hence, the lines *yb* and *t q* will be equal; and if the axle at *cd* is urged forward with an equal force, it will alſo urge the axis of *q s* with an equal force, during half a revolution, or while it moves from *a* to *u*.

But when *d c* has returned from *u* to *d*, the crank will have moved from *r* to *z*, ſomething more than a quadrant, or the crank in that quarter will move a little faſter than the arm, and from *z* to *s* it will not move quite ſo faſt.

Fig. 12.

Given the length of the rod, or diſtance of the centre of the beam or arm from the centre of the crank, and the length of the crank; or, if the ratio of their lengths be given, we may compute the force with which the beam acts upon the crank, in any given poſition. For, ſuppoſe the beam in the poſition A C, and the crank in B D, then the force of the beam will be P C, or rather $\frac{\text{CG}}{\text{PC}}$; the reſiſtance of the crank will be B S, or $\frac{\text{BD}}{\text{BS}}$; hence, in any given poſition of A C, if we can obtain P C and B S, we know the force exerted upon the crank.

The length of the rod BC = AD = 10
The length of the beam AC = rad. = 1 or radius
The angle F C A - - - - - = 30

To find P C and B S.

As

As CB + CA : CB − CA :: co tangent of half ACF : the tangent of half the difference of the angles, viz CBA and CAB, viz. 10 + 1 : 10 − 1 :: 3.732 : 3 0534 = the tangent of 71° 52′, and the angle CAB = 146° 52′, and CBA = 3° 8′.

And as 3° 8′ : 30° :: 1 : 9.147 = AB.

In the triangle ABD, all the ſides are given to find the angles AD = 10

AB =	9.147
BD =	1
Sum	20.147.
Half ſum	10.0735.

From which we find the ſegment SD = .8666.
And the other - - - - - SA = 9.13338.

Alſo the angle EBD is found 36° 9′, when ACF is 30° which proves, as before obſerved, that the crank riſes faſter in this quarter than the beam.

We alſo find PC = .59201, and BS = .49899, and the force exerted on the crank is .8444, but on the deſcending ſide, it is always the ſame, that is, through the whole deſcent equal to 1.

The following table will ſhew the force acting on the crank for different magnitudes of the angle ACF, when the rod is 10 times as long as the crank.

Alſo,

Alſo, the force, with which a long beam, vibrating through a few degrees, acts upon it when in the ſame poſition.

Angle A C F.	Angle P A C.	Angle B D S.	Length P C.	Length B S.	Force on the Crank.	Force of a common beam on the crank.	Rod 20 times length of the crank.
10	12° 21 13	9° 33 51″	21388	.16619	.777	.1736	.790
30	36 13	29 56	.59034	.499	.844	.5	.910
60	70 28	60 59	.9424	.8744	.927	.866	
90	78 34	89 27	.9801	1	1.022	1.000	
	R K C	B R K.					
120	50 26	62 26	.7708	.8864	1.149	.866	1.06
150	24 49	29 17	.4197	.4891	1.165	.5	
170	8 13	9 42	.1429	.1684	1.178	.1736	

The

The inequality of the force upon the crank, in the ſemicircle EDR, is owing to the obliquity of the connecting rod AD or KR; of conſequence the leſs the length of the crank, compared with the length of this rod, the leſs will be the difference of the force, or action of the machine upon the crank, in the two ſemicircles, for whether it moves round one way or the other the difference is the ſame.

In fig. 14, CB = PR = PD = FE the length of the connecting rod, and when the end of the beam is at F the crank will be at E; but, when the beam has paſſed over a quarter of a circle to P, the crank will be found at D, 101° 19′ from E, when the rod is 10 times its length. For as PR : RS :: radius: tangent of 11° 19′ the angle RPS which is equal to the angle DBS which added to 90° gives 101 19′.

When the rod is ſhort compared with the length of the crank, for inſtance, 4 times its length, then when the beam has riſen 30° the force on the crank will be .67; but when the rod was 10 times the length of the crank, the force at the ſame angle was .77; and, as before obſerved, the longer the rod or ſpear, compared with the crank, the nearer we arrive at an equal preſſure Beſides, in this caſe, a rod fixed to a long beam, and only 4 times the length of the crank, would only act with a force equal to .46 imtead of .67.

N. B.

N. B. If the beam vibrates through a ſemicircle, it is poſſible, in ſome circumſtances, it might return the wrong way; to prevent which, it may be proper to make it a little longer than the crank, ſo as to paſs over 150 or 160°, and though its force will not be quite ſo equal as when they are equally long; nevertheleſs it will be much better than the common method.

A different demonſtration.

Fig. 13.

In the triangles EBD and BDC, we have DB common to both, and by ſuppoſition EB = DC. Let BE, BD, and the angle DCF be given, to find the difference of the angles DCF and BEA, when the arm DC has moved from F to D.

Solution. With the angle DCF, are given DO and CO, and of conſequence EO. Then as EO : DO :: radius : tangent of DEO; and as ſine of DEO : DO :: radius : DE, it is alſo evident that the arch SB is equal to DF, but while the point D has paſſed over D , B has paſſed over TB, leſs than the other by the arch ST = the angle DEO.

ON

ON

LATHES.

WORKMEN who turn wood, braſs, iron, &c. in the ſame lathe, find it neceſſary to turn different ſubſtances with different degrees of velocity; hence, a wheel with a number of grooves, of different diameters, correſponding with grooves of different diameters in a pully or whirl upon the ſpindle of the lathe, and in ſuch manner that the band, when ſhifted from groove to groove, ſhall always be equally tight.

PROBLEM.

Fig. 16.

Given $CD = r = .7$ foot.

$CP = d = 2$ feet.

$EFGE = l = 7.1416$ the length of the chord.

$PF = DE = x$.

$CE = r + x$.

3.1416

$3.1416 = p.$
$360 = q.$

Tangent of DCA $= t$, to find x.

As $r : d :: 1 : \frac{d}{r} = 2.8571428 =$ the secant of $69° 31'$ the arch DA, or FL, which, doubled, is equal FQ, the number of degrees on the whirl touched by the cord $= 139°.2' = s$ and $\frac{2psx}{q} = 2.4265369x =$ the length of the cord on the whirl.

And $q - s = n =$ the degrees on the large wheel touched by the cord. For $\overline{2r + 2x} \times p =$ its circumference.

As $q : 2pr + 2px :: n : \frac{2npr + 2npx}{q} =$ the length of cord which touches the wheel.

As $\frac{d}{r} : t :: d : tr =$ PD $=$ EF $=$ the length of cord between the wheel and whirl.

$\frac{2spx}{q} + \frac{2npr + 2npx}{q} + 2tr = l$, the whole length of the cord, from which we find, $x = \frac{ql - 2qrt - 2npr}{2sp + 2np} = \frac{\frac{1}{2}ql - qtr - npr}{pq} = \frac{57.295l - 114.79tr - nr}{360}$

If

If CD $= r$ be taken as below, x and CE will be found as placed in the ſame line.

CD.	X.	CF.	Whirl turns times for the wheel once.
or $r = .4$	.287	.687	2.4
$r = .5$	.228	.728	3.19
$r = .6$	.1714	.7714	4.5
$r = .7$	.1105	.8105	7.33
$r = .8$	.0485	.8485	17.4

The radius of the wheel cannot be much longer than in the laſt line, except the cord is made longer, which it ought to be, if a quick motion is required.

In a wheel lathe turned by the foot, 3 turns of the whirl, for one of the wheel, for iron; 4 of the whirl, for one of the wheel, for braſs; and 8 or 10 for wood, will be found ſufficiently quick.

Let the diſtance of the centres CF be $2.5 = d$ and $l = 9$, all elſe as before; then by taking $r =$ the oppoſite numbers, we ſhall have x as in the following ſcheme.

ON

ON

WHEEL CARRIAGES.

THE following propositions, as they are called, were put into my hands by a gentleman at Lancaster, many years since.

PROPOSITION I.

"Equal powers, whether they be the exertions of muscular motion, or the force of gravity, produce equal effects in moving wheels over equal obstacles, on an horizontal plane, though the diameters of the wheels be unequal.

PROPOSITION II.

But the effects of these powers are not equal in moving wheels up an inclined plane, when the diameters are unequal; the magnitude of the obstacles being encreased by the increase of diameter; but the encrease of diameter does not increase the effect, by proposition 1.

PROPOSITION

PROPOSITION III.

Or, the *tendency* of a wheel to roll down an inclined plane, the *power* requiſite to roll it up, and the *force* neceſſary to keep it in *ſtatu quo*, being all equal to each other, that is, being one and the ſame thing, an encreaſe of the diameter increaſes the perpendicular diſtance of the point of ſupport, or contact, from the horizontal line drawn from the loweſt point of the circumference the wheel, increaſes the tendency to roll down, conſequently increaſes the power neceſſary to roll it up; but by propoſition 1, the effects are only equal, when the ſame perpendicular diſtance of the point of contact from the horizontal line is equal, though the diameters of the wheels are unequal.

PROPOSITION IV.

Or, the magnitude of the obſtacles being encreaſed by the increaſe of diameter in moving a wheel up an inclined plane, and the number of obſtacles being equal, an increaſe of diameter neceſſarily requires an increaſe of power to overcome them, but the effect is not increaſed by increaſing the diameter, by propoſition 1."

Obſervations on the foregoing propoſitions.

Fig. 17.

Let $a\ b$, $q\ b$, be two obſtacles equally high, on the

the horizontal plane BB, one obſtructing the wheel DB, 3 feet high, and the other obſtructing the wheel FB, 6 feet high. And let the height of ab = 6 inches, then will Dn = 2.5 and nq = 1 658, and the angle BDq = 33° 33′ 27″; its ſine = .5527735, its co-ſine 8333 the obſtacle qb, or its equal the plane qd = 33° 33′ 27″. We need not go through Euclid to prove, that when the weight of the carriage is D n; the force to ſupport it againſt the obſtacle qb, when B is juſt raiſed from the horizontal plane BB, is nq; or, as Dn : nq :: co-ſine to the ſine of BDq; in this caſe, as .8333 : .5527 or as 10 : 6.633; hence, it appears that if the weight of the 6 feet wheel is 10, it will require above 6½ to draw it over an obſtacle 6 in high

For the leſſer wheel, as ac : cn :: radius to 66666, the co-ſine of the angle aCB = 48° 11′ 23″, its ſine is .7453564 = co-ſine of 41° 48′ 37″, and as Cn : an :: 1 : 1.1180346.

Hence, the force requiſite to move the 3 feet wheel over the 6 inch obſtacle, is to that required to move the 6 feet wheel over the ſame, as 11.18 to 6.63; for if the weight of the wheel as before is 10, then as 1 : 1.118 :: 10 : 11.18.

PROPOSITION II.

Fig. 18.

In two wheels placed on an inclined plane, the

lines which fall from their centres perpendicular to the horizon, and the lines which go from their centres perpendicular to the inclined plane, which forms the point of contact, make the same angle with each other, whatever be the size of the wheel.

And as the tendency to roll down the plane is as the sine *na* to the radius C*a*; or as *bc* to AC, the same power will support or draw them up the plane.

Had our writer been acquainted with the composition and resolution of forces, he would have known, that if a weight *w* hangs from the end of a prop PF, and is supported by a line PL, whatever may be the position of PL it will require the same force to support the weight *w*, whether PF be long or short. And if PL is parallel to DF, the wheight of the body, and the force to support it, will be as PF to GF; and PF and GF will still have the same ratio. Hence, on the same inclined plane, all wheels of equal weight are supported or drawn up by the same power, not that the power which supports will draw them up, but whatever power will draw up a small wheel will draw up a large wheel of the same weight, our author though unacquainted with mechanics, might, for the expence of sixpence, have convinced himself by experiment.

On

On an horizontal plane a large wheel is much easier drawn over an obstacle than a small one, or the resistance diminishes with the magnitude of the wheel, *therefore the first proposition is false.*

On an inclined plane all wheels of equal weight require the same force to raise them, and if there are obstacles upon the inclined plane, the large wheel has an advantage as upon the horizontal plane, *therefore the 2d 3d and 4th propositions are all erroneous.*

Many gentlemen have supposed that a large wheel has more tendency to roll down a plane than a small one; if there are obstacles upon the plane this is *true*; and if it is *true*, the great wheel would be easier drawn up such a plane.

A

A STREIGHT LINE,

DESCRIBED BY THE END, OR SOME OTHER POINT, OF

A MOVING BEAM.

Fig. 19.

GIVEN ABE, a right angle, and if the end A of the right line AD defcend from A to B, along the line AB, while the end D moves along the line BE, a point C in the middle of the line will defcribe the circle CC.

Demonftration.

When AB = DB, the angle BAD = BDA = BCP = PCD; hence, the lines AC, DC, BC muft be equal.

And when BC is in the pofition BT, CD will be in the pofition TE, and BT being equal to TE, the angles TBC, TEC muft be equal, alfo NTB = twice TBC; hence, NT = BT = CD = TE.

If

If a beam AD has one end ſliding in a groove at D, and is connected or jointed at the middle C, to a guide BC, of half its length, this guide alſo moving on a joint at B, then will ACB, in every poſition of the beam AD, be an iſoſceles triangle, the point C will deſcribe the circle CE, and the point A will move in the ſtreight line ABA.

If AD is 12 feet, and AA 6 feet, the angle CDE will be about $14\frac{1}{2}$ degrees.

Fig. 20.

If the point C is not in the middle of the line AD, it will deſcribe an ellipſis, if between A and C, AB will be part of the tranſvere diameter, but if between C and D, it will be the conjugate.

Demonſtration.

$$\text{Let } \left.\begin{array}{r} AD = 10 \\ AG = 4 \\ BD = 6 \end{array}\right\}$$

$$OB = GD = 6$$

Then $AD^2 - BD^2 = AB^2$; or, $100 - 36 = 64$ and $AB = 8$.

As $AD : AB :: AG : AN$; or, as $10 : 8 :: 4 : 3.2 = AN$

As $AD : BD :: AG : NG$; or, as $10 : 6 :: 4 : 2.4 = NG$ $AB - OB = AO = 2$; and $AN - AO = NO = 1.2$

As $OB : BC^2 :: \overline{2OB - NO} \times NO : NG^2$, a property of the ellipſis.

As

As $6 \times 6 : 4 \times 4 :: \overline{12 - 1.2 \times 1.2}$ $(= 12.96)$ 2.4×2.4.

Or, as $36 : 16 :: 12.96 : 5.76$, the ſquare of the ſemi-ordinate NG. Take the beam AD in any other poſition; for inſtance, let $BP = 8$; then $AD - BD^2 = OB^2 = 36$.

And as $PO : PB :: SO : SR$; as $10 : 8 :: 4 : 3.2 = SR$.

As $PO : BO :: SO : OR$; as $10 : 6 :: 4 : 2.4 = OR$.

And $OB - OR = BR = 3.6$. Then as $OB^2 : BC^2 :: \overline{OB + RB} \times OR : RS^2$, viz. $6 \times 6 : 4 \times 4 :: \overline{6 + 3.6} \times 2.4 : 10.24 = RS^2$; or, $36 : 16 :: 23.04 : 10.24$ the ſquare root of which, is 3.2, the ſame as found above independent of the property of the ellipſe.

Fig. 21.

If the line AD moves with the end A in the line EI, and the end D in the line DI, then will the point C deſcribe the ellipſis qCE, the tranſverſe diameter will be equal to twice CD, for when D is at I, C will be at E. The conjugate will be equal to twice AC, or twice Iq; we have therefore to find the centre of a circle which ſhall paſs through Cqr; which is done by dividing the ſquare of Cp by pq, and to the quotient adding pq, half the ſum will be BC.

To find pq, we have, as EI ſquared is to Iq ſquared,

squared, so is EI + IL × EL : LC squared, the square root of which taken from ɪq, leaves pq.

Or, as DA : AI :: CA : AL, then $\overline{AC}|^2 - \overline{AL}|^2 = \overline{LC}|^2$ and AC—LC $= pq$.

Then as DA : AI :: DC : cp, and $\frac{\overline{cp}|^2}{pq} + pq =$ 2BC.

PROBLEM.

Let AD = 14 = a
AC = 6 = d
AI = 3 = b
EI = CD = 8 = n.

As $a : b :: n : \frac{bn}{a} =$ Cp, as $n^2 : d^2 :: \overline{b - \frac{bn}{a}} \times \overline{n + \frac{bn}{a}} : \frac{d^2b^2a^2 - d^2b^2n^2}{a^2n^2} = \overline{LC}|^2$, and LC taken from AC leaves pq.

In numbers.

As 64 : 36 :: 61.06124440816 : 34.3469497959, the square root of which is 5.860627, which, taken from 6, leaves .139373 = pq.

Otherwise, as 14(a) : 3(b) :: 6(d) : 1.28571427 = AL $\sqrt{6^2 - \overline{1.2857\ \&c.}|^2}$ = 5.860627; which, taken from 6, leaves .139373 = pq, as before.

And as 14 : 3 :: 8 : 1.7142857 = Cp, $\overline{cp}|^2$ = 2.938775, which divided by pq, .139373, quotes 21.085,

21.085, to which add pq, and we have 21.2247, which, divided by 2, gives 10.6123 = BC.

Let the deviation of A, from the line AI, be computed.

Let the point A be moved to L, ſo that C ſhall be at $x = 1$, from p. Then will $\frac{Ey \times \overline{EI + yI} \times \overline{EI}^2}{\overline{EI}^2}$ $= \overline{xy}^2$, the ſquare root of which is 5.95294; which, taken from Iq, leaves .04706.

But in the circle, when Bq is the radius = 10.6123, we have $\overline{Bq + 1} \times \overline{Bq - 1} = 111.62091129$, the ſquare root of which is 10.56503, the coſine of the arch xq, which, taken from the radius BC, leaves .04727, which is more than the ſemi-ordinate xy of the ellipſes by .00021, which, if the dimentions be in feet, will amount to $\frac{252}{100000}$ of an inch, or near one five hundredth part of an inch

The motion of the end D will be equal to the verſed ſine of the angle ADI, in this problem = 325206; hence, it may be more convenient to have it fixed to the end of a bar ND, of ſome feet in length, than to ſlide in a groove, for, though the ſmall arch deſcribed by the end D will deviate a little from a ſtreight line, yet, the error produced thereby will be very ſmall. In this caſe ſuppoſe

pose DN to be three feet, then will the angle described by it be about 6° 13′, and the versed sine of the arch .0014633. And as CD : CA :: .0014633 : 00109747 the motion of A, in the circle AZ. And as IAD : ADI :: .0010974 : .00024061, the deviation of A from the perpendicular; this error is in a contrary direction from that found above, and in part corrects it, but if they both remain, they amount to nothing that could produce any effect, or ever be discovered in practice.

Fig. 22.

PROBLEM.

Let AC = 6
DC = 3
AI = 3

Then as AD : AI :: DC : cp = 1, as 9 : 3 :: 3 : 1.

As AD : AI :: AC : AL = 2, for as 9 : 3 :: 6 : 2.

$\sqrt{\overline{AC^2 - AL^2}} = LC = 5.656$, which taken from AC, leaves .344 = pq.

$\frac{cp^2}{pq} = 2Bq - pq$, and $Bq = BC$, the length of the guide, in numbers.

$\frac{1^2}{.344} = 2.907$ to which, add .344, and we have 3.241, which divided by 2, gives 1.6205 = BC.

PUMPS.

Fig. 23.

LET AP be a pump, P the piſton, B the valve.

Put $a =$ AB, the altitude of the valve above the water.

$b =$ BP, the length of the ſtroke.

$1 =$ the preſſure of the atmoſphere.

$x =$ the height AC, that the water riſes the firſt ſtroke.

$d = 32$ feet, the altitude of a column of water ſupported by the preſſure of the atmoſphere.

Then after the firſt ſtroke the air which was contained in AB, will be contained in CP $= a + b - x$; and its preſſure upon the water will be $\frac{a}{a+b-x}$ of what it was before: or of 1, the whole preſſure of the atmoſphere. The weight of AC, or x, the elevated water will be $\frac{x}{32}$ part of the atmoſphere.

Hence, $\frac{a}{a+b\quad x} + \frac{x}{32} = 1$, the whole preſſure of the atmoſphere. Then $da + ax + bx - x^2 =$ ad

$ad + bd - dx$; by tranſpoſition—$db = x^2 - dx - bx - ax$; putting $2n = d + b + a$, we get $x = n + \sqrt{n^2 - dl}$, Theo. 1.

Example I.

Let $a = 10$ feet.

$b = 1$ foot, then will $a + b + d = 43$ feet $= 2n$.

And $x = 21.5 + \sqrt{462.25 - 32} = .76$ feet or 9 inches nearly.

Example II.

Let $a = 20$.

$b = 1$.

$2n = 53$.

Then $x = 26.5 + \sqrt{702.25 - 32} = .62$ feet $7\frac{1}{2}$ in.

Example III.

Let $a = 28$.

$b = 1$.

$2n = 61$.

Then $x = 30.5 - 29.97 = .53$ feet $= 6\frac{1}{4}$ inches.

For the ſecond ſtroke.

The length of the column of air below the piſton is equal to $a - x$, which put $= m$, and let x equal the height to which the water will riſe at the firſt and ſecond ſtrokes together, then the expreſſion will become $\frac{m}{a+b-x} + \frac{x}{32} = 1$, which cleared of of fractions, becomes $ax + bx + dx - x^2 = da + db - dm$, and by putting $a + b + d = 2n$, we get $x^2 - 2nx = dm - da - db$; from which by compleating the square, we get $x = n + \sqrt{n^2 + dm - da - db}$, Theo. 2.

In example I.

$a - x = 9.24 = m$, and $a + b + d = 43$, and $x = 21.5 - \sqrt{462.25 + 295.68 - 320 - 32} = 1.36$, the afcent at both ftrokes, from which take .76, the afcent at the firft ftroke, and we have .6 foot, or 7.2 inches for the fecond ftroke.

For the third ftroke.

Put $z =$ the altitude to which it will rife at the third ftroke, then 1.36, the altitude to which it rofe at the fecond, being taken from 10 $(a - x,)$ leaves $8.64 = m$. From the theorem $z = 21.5 - \sqrt{462.25 + 276.48 - 320 - 32} = \sqrt{386.73} = 19.66$, then $21\ 5 - 19.66 = 1.82$ feet for the altitude of the water after the third; from which take 1.36, the altitude after the fecond ftroke, and their remains .46 foot, or $5\frac{1}{2}$ inches, for the afcent at the third ftroke.

The fecond or laft theorem is general; by it the afcent of the water may be found after any number of ftrokes.

If the lower part of the tube is narrower than the

the working barrel, then let z be the area of the piſton, and y that of the ſmall pipe, and $\frac{z}{y} \times b$ will be the length of the ſtroke; or the ſpace left by the piſton for the air to expand in; and will be equal to the ſpace that would have been left, had the tube been equally wide, by a ſtroke $= \frac{zb}{y}$; and the theorem will become $x = n + \sqrt{n^2 - \frac{zbd}{y}}$; where $2n = a + d + \frac{zb}{y}$.

$$\text{Let } z = 25.$$
$$y = 8.$$
$$b = 1.$$
$$a = 10.$$
$$d = 32.$$

Then $x = 22.5625 -- 20,2253 = 2.3372$ feet.

To find the weight of water contained in a pump; or, the force with which it preſſes upon the piſton.

PROBLEM.

Given, the diameter of the piſton and the height, to which the water is raiſed, to find its preſſure upon the piſton.

RULE.

Multiply the ſquare of the diameter of the piſton in inches by 7854; and that product, by the height in feet: divide the laſt product by 2.304, and

and the quotient will be the weight upon the piſton in pounds avoirdupoiſe.

Example I.

Given, the diameter 6 inches, the depth 10 feet.

Solution.

$6 \times 6 \times .7854 \times 10 = 282.744$; which divided by 2.304, quotes 122.7 pounds, for the preſſure upon the piſton.

Example II.

Given, the diameter of the pump 5 in. and the depth 40 feet, required the preſſure upon the piſton.

Solution.

$5 \times 5 \times 40 \times .7854 = 785.4$, which divided by 2.304, gives 340.8 pounds, the preſſure ſought.

The following method is a little ſhorter.

RULE II.

Multiply the ſquare of the diameter of the piſton in inches, by the depth in feet, and again by .3408, the laſt product is the preſſure in pounds, &c.

Example II. wrought by Rule II.

$5 \times 5 \times 40 \times .3408 = 340.8$, the preſſure as before.

N. B. In making theſe computations, the depth muſt always be meaſured from the ſurface of the water, which is raiſed to the place of delivery, the neglect of which, has cauſed many to commit great errors.

RULES

RULES AND OBSERVATIONS

RESPECTING THE FORM AND STRENGTH OF

BEAMS OF WOOD AND IRON,

FOR SUPPORTING WEIGHTS, WORKING ENGINES, &c.

IF the materials of which different beams are made, be equally good, the comparative ſtrength under any regular form may eaſily be inveſtigated. But we find by experiment that the ſame kind of wood, and of the ſame form and dimenſions, will break with very different weights; or, one piece is much ſtronger than another, not only cut out of the ſame tree, but out of the ſame rod; or, a piece of a given length, planed equally thick, and cut in two or three pieces, theſe pieces will be broken with different weights. Iron alſo varies in ſtrength, and not only from different furnaces, but from the ſame furnace, and the ſame melting; but this ſeems to be owing to ſome imperfection in the caſting, and in general iron is much more uniform than wood. The reſiſtance which any

 beam

beam of wood or iron affords, will be as the ſum of the products of all the fibres, between the top and bottom, multiplied by their reſpective diſtances from the top. For if $a =$ length, $b =$ breadth and $z =$ depth, we ſhall have $z \times z$, and divided by $\frac{a}{2}$; the fluent of $zz = \frac{z^2}{2}$; hence, $\frac{bz^2}{2} =$ the whole reſiſtance, which when the weight is ſuſpended from the middle of the beam, muſt be divided by half the length, or, by $\frac{a}{2}$, which will be equal to $\frac{bz^2}{a}$; which expreſſes the ſtrength of the beam. From which we have the following

RULE.

Multiply the breadth in inches by the ſquare of the depth in inches, and divide that product by the length in inches, the quotient is a fraction, or whole number, &c, which expreſſes the comparative ſtrength of the beam.

The dimenſions may be taken in feet, or the breadth and depth in inches, and the length in feet, but to compare one piece with another they muſt all be taken in the ſame manner. From a great number of experiments which I have made on the ſtrength of wood, and that on pieces of various lengths and breadths, &c. I found that the worſt or weakeſt piece of dry heart of oak 1 inch ſquare

ſquare and 1 foot long, did bear 660 pounds, though much bended, and two pounds more broke it. The ſtrongeſt piece I have tried of the ſame dimenſions, broke with 974 pounds.

The worſt piece of deal I have tried, bore 460 pounds, but broke with 4 more. The beſt piece bore 690 pounds, but broke with a little more. Theſe pieces were 1 inch ſquare, and 1 foot long.

Example I.

Given a piece of oak 6 inches ſquare, and 8 feet in length, to find what weight ſuſpended from the middle will break it.

Solution.

In the worſt piece of oak 1 inch ſquare, and 12 inches long, the ſtrength is 1 ſquared, viz. the depth ſquared and multiplied by the breadth, and divided by the length, which is $\frac{1}{12}$. In the given piece we have 6 the depth ſquared equal 36, which multiplied by the breadth (6) gives 216, which divided by the length 96, gives $\frac{216}{96}$, or $\frac{9}{4}$; hence as $\frac{1}{12}$ is to 660 pounds, ſo is $\frac{9}{4}$ to 17820 pounds.

Example II.

Let the piece be 4 inches broad, 6.364 inches deep, and 6 feet long; what weight at the centre will break it?

 6.364

$6.364 \times 6.364 \times 4 = 162$, which divided by 72, the length, gives 2.25 for the ſtrength.

And as $\frac{1}{12}$ is to 660, ſo is 2.25 to 17820; hence, this piece is equally ſtrong with the laſt, and its ſtrength is expreſſed by the ſame number, for $\frac{9}{4} =$ 2 25.

From the above, we may compute the following weights, which placed oppoſite to the fraction or whole number, which is obtained by the rule before given, and the dimenſions taken in inches; in the ſecond column, in feet; in the third, the breadth and depth in inches, and the length in feet.

The ſquare of the depth multiplied by the breadth and divided by the length.	Weight in pounds which will nearly break it, 1ſt.	Dimenſions taken in feet, 2d.	Weight in pounds.	Breadth & depth in inches.	Length in feet, 3d.
$\frac{1}{12}$	660	$\frac{1}{1728}$	660	1	660
$\frac{1}{10}$	792	$\frac{1}{80}$	14256	$\frac{4}{3}$	880
$\frac{1}{8}$	990	$\frac{1}{60}$	19008	$\frac{5}{2}$	1650
$\frac{1}{6}$	1320	$\frac{1}{40}$	28512	3	1980
$\frac{1}{4}$	1980	$\frac{1}{30}$	38016	4	2640
$\frac{1}{2}$	3960	$\frac{1}{20}$	57024	5	3300
1	7920	$\frac{1}{10}$	114048	6	3960
2	15940	1	1140480	8	5280
3	23700			10	6600
4	31680				

Ex.

Example,

Given, the length of an oak beam 16 feet, breadth 15 inches, and depth 18 inches, required its ſtrength; or, the weight which ſuſpended from the middle, will nearly break it.

1. Let the dimenſions be taken in inches, and we have $\frac{18^2 \times 15}{192} = 25.3125$; then from the firſt column of the table we ſay, as $\frac{1}{12}$ is to 660 pounds ſo is 25 3125 to 200475 pounds, the anſwer.

2. Let the dimentions be taken in feet, and we have $\frac{1.5 \times 1.5 \times 1.25}{16} = .1757 = \frac{45}{256}$; and from the ſecond column in the table we may ſay, as $\frac{1}{1728}$ is to 660 pounds, ſo is $\frac{45}{256}$ to 200475.

3. The breadth and depth in inches, and the length in feet, and we ſhall have the breadth multiplied by the ſquare of the depth, equal 4860, which, divided by 16, gives $\frac{1215}{4}$ or 303.75, and as in the third column, as 1 is to 660 pounds, ſo is 303.75 to 200475 pounds, the anſwer.

A beam of the above dimenſions is commonly uſed for working a ſteam engine, the cylinder of which is from 20 to 24 inches diameter, ſuppoſe

22 inches, then the greateſt preſſure that can poſſibly act upon the beam will not exceed 10,000 pounds; hence, the beam would require above 20 times the force of the engine to break it, nevertheleſs if it was much weaker, the engine might bend it, and in time break it.

Suppoſe we take the above beam for a ſtandard, or conclude that every beam ought to be able to bear 20 times as much as it is employed to do. Then what muſt be the dimenſions of a beam 20 feet long, to work a cylinder 36 inches diameter.

Solution.

The weight ſuſpended from the ends of the beam will be as the ſquares of the diameters of the cylinders, viz. as 22 ſquared and 36 ſquared, or as 484 to 1296, the laſt divided by the firſt, gives 2.6777, or $\frac{324}{121}$; ſo that the ſtrength of the new beam muſt be 2.6777 times as great as the other. The ſtrength of the firſt, when the breadth and depth are taken in inches, and the length in feet is expreſſed by $\frac{1215}{4}$ which multiplied by $\frac{324}{121}$, gives $\frac{393660}{484}$; equal 813.34 for the ſtrength. If no regard is paid to the ratio of the breadth and depth, the problem is ſimply anſwered by aſſuming the breadth what we pleaſe, ſuppoſe 18 inches, and

z =

z = the depth we ſhall have $\frac{18z^2}{20} = 813.34$; and z^2 will equal $\frac{813.34 \times 20}{18} = 903.71$; the ſquare root of which is $= z$, or the depth $= 30$ inches nearly.

Otherwiſe, let the depth be taken at 27 inches and let b = breadth, then our theorem will become $\frac{b \times 27^2}{20} = 813.34$; or, $b = \frac{813.34 \times 20}{729} = 22.31$ inches.

In words.

RULE.

Multiply the expreſſion for the ſtrength, by the length of the beam in feet, and divide that by the ſquare of the depth in inches, the quotient will be the breadth in inches.

N B. Though the above two beams are equally ſtrong, yet the ſecond contains about $8\frac{1}{2}$ feet more wood than the firſt.

PROBLEM II.

Let it be required to make a beam equally ſtrong with the laſt, and of the ſame length, but that the ratio of the breadth to the depth be as 2 to 3, or in any other proportion? Let a = length in feet, b = breadth, z = depth, and $s = 813.34$, the expreſſion for the ſtrength. Then

Then will $\frac{bz^2}{a} = 813.34$; and $z^2 = \frac{813.34 \times a}{b}$; alſo by the problem, as $2 : 3 :: b : z$; hence, $2z = 3b$; and $z = \frac{3b}{2}$, which ſquared, gives $z^2 = \frac{9b^2}{4}$; alſo $\frac{sa}{b} = z^2 = \frac{9b^2}{4}$; from whence we get $b^3 = \frac{4sa}{9} = \frac{4 \times 813.34 \times 20}{9} = 7229.7$ the cube root of which, is 19.337, the breadth in inches, and as $2 : 3 :: 19.337 : 29.005$, the depth in inches.

If m is to n as the breadth to the depth, we ſhall have the following general theorem, viz $b^3 = \frac{m^2 sa}{n^2}$

PROBLEM III.

Required to make a beam 24 feet long, to work a cylinder 24 inches diameter, and that the breadth be to the depth, as 3 to 7.

To find an expreſſion for the ſtrength of the beam, we may ſay from the laſt problem, as 484, the ſquare of the diameter of the cylinder, is to the expreſſion for the ſtrength of its beam, ſo is 576, the ſquare of the diameter of the preſent cylinder, to a number which will expreſs its ſtrength, viz. as $484 : \frac{1215}{4} :: 576 : 361.487$, the number required.

But when the length is taken in feet, and the breadth and depth in inches, we know that an oak beam,

beam, which will juſt break with 660 pounds, has its ſtrength expreſſed by 1, and if we wiſh to have a beam which will bear 20 times as much as it is intended to load it with, we may take $\frac{1}{20}$ part of the load, viz. $\frac{1}{20}$ of 660, or 33, and ſay, as 33 : 1 :: twice the area of the intended cylinder multiplied by 14, to the ſtrength of the beam, in this caſe, as 33 : 1 :: 12672 : 384, the ſtrength. But this is much more than the real load of a common 2 feet cylinder, if it ſhould amount to even 10 pounds per inch, the whole would be only 8640 pounds, and as 33 : 1 :: 8640 : 261.8 the ſtrength; let a beam be conſtructed by both expreſſions, viz by 384 and 262. In the firſt caſe, by the general theorem, $b^3 = \frac{m^2 sa}{n^2}$; in the problem, $m' = 3$, and $n = 7$, alſo $a = 44$, and $s = 384$, therefore $\frac{9 \times 384 \times 24}{49} = b^3 = 1692.7$, the cube root of which is the breadth $= 11.918$ inches; and as 3 : 7 :: 11.918 : 27.808 inches, the depth. I have known a beam of deal of this ſize, uſed for a two feet cylinder, but in a few years it was much bended, though there was no danger of its breaking.

Secondly, when the ſtrength is expreſſed by 262, we have every thing the ſame as before, except $s = 262$, therefore $\frac{9 \times 262 \times 24}{49} = b^3 = 1155$, and the breadth $= 10.492$ inches, and as $3 : 7 :: 10.492 : 24.48$ inches, the depth.

But ſuppoſe the engineer wiſhes it to ſupport any greater weight, for inſtance, 25 times its common load, he may divide 660 by 25, and the quotient will be 26.4; then as $26.4 : 1 :: 8640 : 327.2$ the expreſſion for the ſtrength, and $\frac{327.2 \times 24 \times 9}{49} = b^3 = 1442.351$; its cube root is $11.29 = b$. And as $3 : 7 :: 11.29 : 26.34$, the depth.

By the above proceſs, we find the ſtrength of a beam to work a 12 inch cylinder expreſſed by 96, from which we have the following

RULE.

Square the diameter of the cylinder in feet, and multiply the product by 96; the laſt product will expreſs the ſtrength.

N. B. The depth and breadth are taken in inches, and the length in feet, which I think is the moſt convenient in general. The length of the beam will make no difference, for the dimenſions will turn out ſo as to make it equally ſtrong of any length, for example.

PROB.

PROBLEM IV.

Required a beam 16 feet long, to work a cylinder of 30 inches diameter, and that the breadth be to the depth, as 3 to 5.

Solution.

2.5 feet ſquared, is 6.25; which by the rule, multiplied by 96, gives 600 for the ſtrength.

Then by the theorem, $b^3 = \frac{m^2 sa}{n^2}$; $m^2 = 9$; $n^2 = 25$; $s = 600$; $a = 16$, therefore $\frac{9 \times 600 \times 16}{25} = b^3 = 3456$; the cube root of which, is 15.119 inches, and as 3 is to 5, ſo is the breadth 15.119 to the depth 25.198 inches.

Let it be required to make a beam for the ſame cylinder of 24 feet, then will $\frac{9 \times 600 \times 24}{25} = b^3 =$ the breadth $= 5184$, the cube root of which, is 17.307, from which we find the depth 28.845 inches. This is equally ſtrong with the firſt, but *contains much more wood.*

PROBLEM V.

Required the breadth and depth of a beam 24 feet long, to work a cylinder 4 feet in diameter; the breadth being to the depth as 2 to 3. Firſt, the diameter ſquared and multiplied by 96, gives

1536 for the ſtrength, then $\frac{4 \times 1536 \times 24}{9} = b_3 =$ 16384, its cube root is 25 398 inches, the breadth, and as 2 : 3 :: 25.398 : 38.097 inches, the depth.

PROBLEM VI.

Required a deal beam 16 feet long, to work a cylinder of 1 foot diameter, the breadth to the depth, as 3 to 5.

Solution.

If we take the preſſure at 14 pounds per inch, the weightupon the centre will be about 3168 pounds, alſo the worſt piece of fir, one inch ſquare, and one foot long, bore 460 pounds, and broke with 7 pounds more; hence, if we take $\frac{1}{20}$ part of 460, viz. 23 pounds, and ſay, as 23 : 1 (the expreſſion for its ſtrength,) ſo is 3168 : 137.7 the ſtrength. Then from the theorem we multiply the length, the ſtrength, and the ſquare of the ratio of the breadth together, and divide by the ſquare of the ratio of the depth, viz. $\frac{16 \times 137.7 \times 9}{25}$ is 973.152, the cube root of which, is 9.9 inches for the breadth, then as 3 : 5 :: 9.9 : 16.5 inches, the depth.

Hence, if the ſquare of the diameter of any cylinder in feet, is multiplied by 138, for the ſtrength

ſtrength of a deal beam, it will, without breaking, ſupport more than 20 times as much weight as the engine can ever exert upon it.

PROBLEM VII.

Required a deal beam 24 feet long, to work a 4 feet cylinder, breadth to the depth, as 2 to 3.

Solution.

The ſquare of the diameter multiplied by 138, gives 2208 for the ſtrength, which multiplied by the length, and by the ſquare of the ratio of the breadth, and divide by the ſquare of the ratio of the depth $\frac{2208 \times 24 \times 4}{9}$, gives $23552 = b^3$; its cube root is $28.642 = b$, the breadth, and as 2 : 3 :: 28.642 : 42.963 the depth in inches. This beam is ſuppoſed to be equally ſtrong with the oak beam, in problem 5, but contains 33 feet more wood.

PROBLEM VIII.

Required to make a deal beam to work a 6 feet cylinder, length 24 feet, and breadth to depth, as 3 to 5.

Solution.

The ſquare of the diameter 36 multiplied by 138, gives 4968 for the ſtrength, which, multiplied by 24, the length, and again by 9, the ſquare of the ratio of the breadth, and divided by 25, the ſquare

ſquare of the ratio of the depth, gives 42923.52 $= b^3$; hence, b is found equal to 35.013 inches, and the depth equal 58.355. N. B. This beam contains 340 feet of wood. If we make a ſquare one of equal ſtrength, $b^3 = sa = 119232$, and a ſide will be 49.218 inches; its ſolidity will be 403 feet, or 63 more than the other.

PROBLEM IX.

To determine the form of the ſtrongeſt beam that can be cut out of a round piece of timber of a given diameter.

Let $a =$ diameter, $z =$ depth, and $b =$ breadth, Then will $a^2 - z^2 = b^2$, and $b = \sqrt{a^2 - z^2}$. But $z^2 b$ gives the ſtrength of the beam, and muſt be a maximum, viz. $z^2 \sqrt{a^2 - z^2} =$ a maximum, its fluxion is $4a^2 z^3 \dot{z} - 6z^5 \dot{z} = 0$, from which z is found equal to $2a\sqrt{\frac{1}{6}}$, viz. $2a \times .408248$; and $b = \sqrt{a^2 - z^2}$.

Example.

Given, the diameter 3 feet, to determine the form.

RULE.

Multiply twice the diameter by .408248, *and we get* 2.449488 *for the depth. For the breadth we have* $\sqrt{a^2 - z^2} = \sqrt{9 - 5.999987} = 1.732$; but to make the proceſs eaſier, *we may ſay as* 17 *is to* 12 *ſo is the depth to the breadth*, which will be ſuffi-ciently

ciently near in practice. The largest square beam that could be cut out of the same tree would have a side of the beam expressed by $a\sqrt{\frac{1}{2}}$, or the diameter multiplied by 7071; that is, equal 2.1213 the side of the square. The first would have its strength expressed by 17936, and its solidity by 423. The square beam has its strength expressed by 16423, and its cubic feet by 449.

N. B. I have not taken any notice of the tendency which the beam has to break by its own weight, but it may be observed that half the weight of a beam equally thick, is acting at the centre to break or bend it; for instance, in problem 8, in the first beam, the stress arising from its own weight would be 5780 pounds added to that of the engine.

Experiments on the strength of oak.

These experiments have been made on pieces of one inch square, and one foot long, and from that size to 2 inches square, and five feet long, to mention all the trials would take up time and room to no great purpose. It may be sufficient to observe, that when the computations made on the different pieces, and applied to pieces 1 inch square, and 1 foot long, that the worst would bear 660 pounds, and the best not more than 974. From similar experiments on deal of the same dimentions

mentions, I found the worſt which I uſed would juſt break with 460, and the ſtrongeſt with 690 pounds.

But in all the computations, I have taken the worſt pieces to compute from, and at the ſame time have made them to bear 20 times as much as the load they have to ſupport; not taking notice of their own weight, which would have made the proceſs much more troubleſome.

PROBLEM X.

Required the length of a piece of oak one inch ſquare, ſo that it may juſt break with its own weight.

Solution.

Let $x =$ the length in feet, and one foot in length weigh $\frac{4}{10}$ of a pound. Then as $1 : 660 :: \frac{1}{x} : \frac{660}{x}$ the weight which will break it, but the bar only acts with half its own weight at the centre, therefore $\frac{4x}{10}$ the weight of the bar, muſt be equal to $\frac{660}{x} \times 2$; viz. $\frac{1320}{x} = \frac{4x}{10}$; or, $13200 = 4x^2$, and $x = \sqrt{3300} = 57 \cdot 44$ feet.

Experiments on the ſtrength of caſt iron.

Of late caſt iron has been uſed in various caſes, in

in place of ſtone or wood, as in bridges, engine beams, pillars, rail ways, or roads, &c. and is ſtill likely to be more in uſe. The following experiments, were given to me by Meſſrs. Reynolds, of Ketley, at the ſame time requeſting me to make them as public as I could, for the advantage of others.

Experiments on the ſtrength of caſt iron, tried at Ketley, March, 1795.

N. B. The different bars were all caſt at one time ou of the ſame air furnace, and the iron was very ſoft ſo as to cut or file eaſily.

Experiment I.

Two bars of caſt iron, one inch ſquare, and exactly 3 feet long, were placed upon an horizontal bar ſo as to meet in a cap at the top, from which was ſuſpended a ſcale; theſe bars made each an angle of 45° with the baſe plate, and of conſequence at the top ſo as to form an angle of 90°, from this cap was ſuſpended a weight of 7 tons, which was left for 16 hours, when the bars were a little bent, and but very little.

Experiment II.

Two more bars, of the ſame length and thickneſs, were placed in a ſimilar manner, making an angle of 22½° with the baſe plate; theſe bore 4

N tons

tons upon the ſcale; a little more weight broke one of them, which was obſerved to be a little crooked when firſt put up. In this caſe the preſſure would be as the ſines of the angles of elivation, viz. as 3826 to 7071; and as 3826 : 4 tons :: 7071 : 7.6 tons, that is if the ſecond bars broke with 4 tons, the firſt ought to have taken 7.6 tons to break them, and as they were not broken it is likely that would, if tried, have been the caſe.

Experiment III.

Another bar was placed horizontally upon two ſupporters, exactly 3 feet diſtant, it bore 6 cwt. 3 qrs. but broke when a little more was added.

Experiment IV.

The ſame experiment repeated, with the ſame reſult.

Experiment V.

The bearings were 2 feet 6 inches apart, the bar bore 9 cwt. and broke. This was perceptably bent with 1 cwt. but bore two ſafely. Three more experiments were tried the next day with the priſms 3 feet diſtant; the average reſult was 6 cwt. 2 qrs. $7\frac{1}{3}$ pounds.

Experiments tried at Colebrookdale, on curved bars or ribs of caſt iron, April, 1795.

Rib 29 feet 6 inches ſpan, and 11 inches high in

in the centre, it ſupported 99 cwt. 1 qr 14 lbs it ſunk in the middle $3\frac{7}{8}$, and roſe again $\frac{3}{4}$ when the weight was removed. The ſame rib was afterwards tried without abutments, and broke with 55 cwt. 0 qrs. 14 pounds.

Experiment VI.

Rib 29 feet 3 inches in ſpan, a ſegment of a circle 3 feet high in the centre, it ſupported 100 cwt. 1 qr. 14 lbs. and ſunk $1\frac{3}{16}'$ in the middle. The ſame rib was afterwards tried without abutments, and broke with 64 cwt. 1 qr. 14 pounds.

N. B. The thickneſs of theſe ribs is not mentioned, but the experiments ſhew that they are much ſtronger with abutments; as little more than half the weight which they ſupport breaks them when the abutments are removed.

The following experiments on caſt iron, I made at Meſſrs. Aydon and Elwell's foundry, at Wake-field, the iron came from their furnace at Shelf, near Bradford, and was caſt from the air furnace; the bars one inch ſquare, and the props exactly one yard diſtant, one yard in length weighs exactly 9 pounds, or one was about half an ounce leſs, and another a very little more; they all bended about one inch before they broke.

 1. The

1. The firſt bar broke with - - 963 pounds.
2. Bar broke with - - - - - - 958 pounds.
3. Bar broke with - - - - - - 994 pounds.
4. Bar made from the cupola, broke with - - - - - - - - - 864 pounds.
5. Bar equally thick in the middle, but the ends formed into a parabola and weighed 6 lbs. 3 ozs. broke with - - - - - - - - - 874 pounds.

Other experiments were made by giving the ſame quantity of iron a different form, (ſee fig. 1 in the plate of caſt iron, or plate 4.) The top and bottom of this beam were each 1 inch broad, and half an inch thick, till they joined at *a* and *b*, where they were 1 inch ſquare; the piece *c*, *d*, in the middle from which the weight was ſuſpended, increaſed the weight of this to 10½ pounds. The length from priſm to priſm, viz. from prop to prop, was an exact yard; the depth in the middle from top to bottom, 4½ inches. The firſt piece bore 29 cwt. 20 lbs. and broke with a little more. A ſecond piece bore 23 cwt. 1 qr. but broke with another half cwt.

A ſecond form is repreſented in fig. 2, where the bar *acb* is 1 inch broad, and $\frac{6}{10}$ inch deep, the bar *adb* is alſo 1 inch broad, and $\frac{4}{10}$ inch thick, ſo that it contains no more iron than the ſtreight bar, except

except the piece at *c*, from which the weight was suspended; the weight of these was 10 pounds each, the depth *c*, *d*, 4½ inches, but there was no connection betwixt the two, except at the ends where they were in one piece.

Trial 1. 40 cwt. 2 qrs. 1 lb. broke the upper piece at *c*, and the lower at the props.

Trial 2. 48 cwt. 2 qrs. 7 lbs. broke this in the same manner as the other. A gentleman present wished to have the upper and lower part connected as at the doted lines; one was cast in this form, and broke with 31 cwt. 2 qrs. another bore above 40 cwt.

Another beam of the same length and depth at the centre, but in the form of a parabola, and weighed 10¾ pounds, the flat part of the beam was ¼ of an inch thick, and was surrounded by a moulding ¼ of an inch thick, and on the outside 1 inch broad; first trial broke with 50 cwt. 3 qrs. 25 lbs. a second piece or beam broke with 44 cwt. 3 qrs. but on examining the fracture, it was full of pores at the gate, or place where the metal entered the mould, fig. 3d.

From the above experiments it appears that cast iron is from 3 to 4½ stronger than oak of the same dimensions, and from 5 to 6½ times stronger than deal.

Iron

Iron is much more uniform in its ſtrength than wood, yet it appears that there is ſome difference in different kinds of ore or iron ſtone; there is alſo a difference from the ſame furnace, perhaps owing to the degree of heat which it has when poured into the mould. If we take iron upon an average to be 4 times as ſtrong as oak, and $5\frac{1}{2}$ as ſtrong as deal, we may proceed to make compariſon betwixt wood and iron, in reſpect of magnitude, weight, expence, &c.

It is proved by the experiments that the worſt or rather weakeſt caſt iron 1 inch ſquare, and 3 feet long, will break with about 730 pounds, and as $\frac{1}{3}$ (viz. the breadth and ſquare of the depth multiplied together, and divided by the length in feet) is to 730 pounds ſo is $\frac{1}{1}$ or 1 to 2190 pounds, the weight which would break a bar 1 inch ſquare, and 1 foot long. But I have computed the beams of wood to bear 20 times as much as the intended load. Let the iron be made to ſupport 6 times the weight, which I preſume, from obſervation, will be ſufficient to keep them from bending, or vibrating. But if the engineer thinks differently he may make them of what ſtrength he pleaſes by the following.

RULE.

Divide 2190 *by the number of times you wiſh to encreaſe*

encreaſe its ſtrength above the load, and ſay as the quotient is to 1, *ſo is the load of your new beam to the number which expreſſes its ſtrength.*

For example, the load upon a 12 inch cylinder will be about 3168 pounds. One ſixth part of 2190 will be 365, and 3168 divided by 365, quotes 8.6; which expreſſes the ſtrength for caſt iron.

PROBLEM XI.

Let it be required to make a caſt iron beam for a 12 inch cylinder, ſo that the breadth may be to the depth as 1 to 6, and the length 14 feet. Here we have $a = 14$; $m = 1$; $n = 6$; $s = 8.6$; then $b^3 = \frac{m\,sa}{n^2} = \frac{120.4}{36} = 3.344$, and $b = 1.4954$ inches, which multiplied by 6, gives 8.9724, the depth.

PROBLEM XII.

Required the dimenſions of a beam to work a cylinder 4 feet diameter, and the breadth to the depth as 2 to 3, the length 24 feet, as in problem 5.

Solution.

By multiplying the ſquare of the diameter of the cylinder, viz. 16 by 8.6, we get 137.6, which expreſſes the ſtrength. Then by the theorem, we have

have $\frac{137.6 \times 24 \times 4}{9} = 1467.4 = b^3$, from which we find $b = 11.363$, and the depth $= 17.044$ inches. The weight of this beam will be about 6 ton 4 cwt. In this caſe the beam is ſuppoſed to be of the ſame breadth and depth through the whole length, but by caſting it in the form of a parabola it would be equally ſtrong, and one third of the weight diminiſhed, or it would be reduced to about 4 tons 3 cwt.

Note.

As iron can be made or caſt in any ſhape at pleaſure, beams may be made equally ſtrong, and at the ſame time much lighter than this: for example,

PROBLEM XIII.

Let it be required to make a beam for the engine in the laſt problem, and of the ſame length, but let the breadth be to the depth as 1 to 18.

Solution.

By the laſt problem we have 137.6 for the ſtrength, which multiplied by 24, gives 3302.4, which divided by 18 ſquared, viz. 324, quotes 10.19 for the cube of the breadth, the cube root of 10.19 is 2.168 for the breadth or thickneſs, which multiplied by 18, gives 39.024 for the depth.

depth. The weight of this is, if of uniform dimenſions through the whole length, about 2 tons 14 cwt. but by caſting it in the form of a parabola, it will be reduced to about 1 ton 16 cwt. Theſe beams will coſt about 15 or 16 pounds per ton; hence, this beam ready for hanging would not amount to 30 pounds. In problem 5, if we could purchaſe a piece of wood of the length required, it muſt contain at leaſt 216 feet in the round, which at 6 ſhillings per foot (but much is ſold for 7 and 8 ſhillings) would amount to 64 pounds 16 ſhillings, after which, there would be the ſawing, hewing, bolting, ſcrewing, &c. which would add much more to the expence.

N. B. In forming theſe parabolic beams, when the thickneſs is little compared with the depth, it may be well to put a moulding round them, 3 or 4 inches broader than the thickneſs of the beam, by which any twiſt to a ſide will be guarded againſt.

To form a parabolic beam obſerve the following proceſs; ſuppoſe the beam is to be 24 feet long from centre to centre, add about a foot at each end, to contain the centres, and then ſay, as 13 feet, half the length of the beam, is to the ſquare of the depth at the middle, ſo is any other diſtance from the end to the ſquare of the depth at that diſtance.

Example.

				Square roots or depth of the beam.
As 13 feet is	1531 ſo is	12 to	1404	37.4
13	1531	11	1287	35.8
13	1531	10	1170	34.2
13	1531	9	1053	32.4
13	1531	8	936	30.5
13	1531	7	819	28.6
13	1531	6	702	26.4
13	1531	5	585	24.1
13	1531	4	468	21.6
13	1531	3	351	18.7
13	1531	2	234	15.3
13	1531	1	117	10.8
13	1531	5	58.5	7 6

The ſecond number in the ratio, 1531, is the ſquare of 39, the depth of the beam at the middle. And the numbers in the laſt column are the breadths of the parabolic beam at every foot from the centre to the end. To lay this down upon wood, paper, &c. draw a ſtreight line, and divide it into 13 equal parts, at the end ſet off 19 5 each way, for the breadth at the centre, then take half of 37.4 in your compaſſes, and ſet it each way for the breadth at the next diviſion, and ſo on till the end. And by drawing a curve through the points, we get the form of one half of the beam,

beam, to which the other half will be ſimilar, (ſee fig. 4, plate 3.)

N. B. The above computations may be made ſufficiently near the truth by a good ſliding rule. For if we ſet half the length of the beam on the ſlide, to the depth at the middle on a double radius, then againſt any diſtance from the end on the ſlide, we have the depth at that diſtance on the double radius; which is all done at one ſetting of the ſlide.

Suppoſe we want a beam 20 feet long, and 30 inches deep at the middle, then adding 1 foot to each end, we ſet 11 on the ſlide to 30 on the rule, and againſt 10 we have 28.6; 9 gives 27.15; 8 gives 25.6; 7 gives 23.9; 6 gives 22.15; 5 gives 20.2; 4 gives 18.08; 3 gives 15.65; 2 gives 12.8; 1 gives 90.4; ½ gives 6.4. Theſe breadths are taken from the ſliding rule without any correction, and will be found near enough to the truth in practice.

Should it be required to make a caſt iron beam to ſupport 10 times the intended load, and that the breadth be to the depth as 1 to 16, and the length 24 feet, as in problem 5 and 13.

In this caſe we have 2190 divided by 10, which gives 219, by which we are to divide 3168, and we get 14.5, (ſee problem 10) by which multiply

16, the ſquare of the diameter of the cylinder, and we have 232 for the ſtrength of this beam, and by the *theorem* $\frac{232 \times 24}{256}$ gives 21.75, for the cube of the breadth, from which the breadth is found equal to 2.7892 inches, and the depth equal 44.637 inches. The weight of this beam, if of equal thickneſs from end to end, would be about 4 tons, but formed into a parabola will be reduced to about 2 tons 14 cwt. and might coſt from 40 to 44 pounds.

If the beam is to be made to ſupport 20 times its intended load, then we muſt multiply the ſquare of the diameter of the cylinder by 29, (ſee problem 10.)

Let one of this ſtrength be made for the ſame engine. Firſt we have $29 \times 16 = 464$, the ſtrength, which multiplied by 24, gives 11136, which divided by 16 ſquared, quotes 39.6 the cube of the breadth, from which we find the breadth 3.4, and the depth 54.5 inches. The weight, when in its proper ſhape, would be about 4 tons.

This beam ready for hanging would be ſtill cheaper than wood, beſides the expence of forming, &c. which could not be leſs than 30 or 40 pounds, ſome ſay much more, but of this the practical engineer will make a much better eſtimate, than I can pretend to do.

PROB.

PROBLEM XIV.

Given, a caſt iron beam 22 feet long, and 28 inches deep, weight 29 cwt. 2 qrs. the form a parabola, to find the thickneſs and ſtrength compared with the load; it works a cylinder of 30 inches diameter.

Solution.

The length in inches multiplied by the depth at the middle is 7392, two thirds of which, is the area of the ſide of the beam, equal 4928 ſquare inches. The weight of the beam is 3304 pounds, and by allowing 4 cubic inches of iron to 1 pound, we have $\frac{4928 \times b}{4} = 3304$; and b the thickneſs will $= \frac{4 \times 3304}{4928} = 2.68$ inches. Next, to find the ſtrength, we have $\frac{28 \times 28 \times 2.68}{22} = 95.5$; then (ſee problem 10) as 1 : 2190 pounds :: 95.5 : 209145 pounds, the weight which would break it. The load of the cylinder will be about 19800 pounds, by which if we divide 209145, we get 10.56, viz. the beam would require above 10 times as much weight to break it as the engine can exert upon it.

PROBLEM XV.

Another caſt iron beam of the ſame length and depth as the laſt, but weighed 48 cwt. is employed

ed to work a 30 inch cylinder, what is its ſtrength?

Solution.

The weight of the beam in pounds is 5376, which multiplied by 4 and divided by 4928, gives 4.36 inches for the thickneſs, which multiplied by the ſquare of the depth, and divided by the length, gives 155.37, for the ſtrength. Then as 1:2190 :: 155.37 : 340260 pounds, the weight which would break it, which divided by the load of the engine 19800, gives 17.18, by which, it appears that it would require more than 17 times the force of the engine to break it. Beſides the engine is ſuppoſed to be loaded with 14 pounds to the inch, but in a common engine it will not be above half as much in general, and as both the above beams are working common engines, the firſt may be conſidered as 20, and the laſt as 34 times ſtronger than that which would juſt break with the power of the engine.

PROBLEM XVI.

A beam 16 feet long, 16 inches deep, and 2 inches thick, has been employed to work a cylinder of 21 inches diameter for ſome years; how much more than the force of the engine would it ſupport?

Solution.

The ſquare of the depth, multiplied by the breadth,

breadth, and divided by the length, is 25.7 for the ſtrength; which multiplied by 2190, gives 56283 pounds, the weight that would break it. The full computed preſſure of a 20 inch cylinder would not exceed 9702 pounds, by which divide the ſtrength, and we have 5.8, ſo that it would bear near 6 times as much weight as the engine can load it with. But, had the ſtrongeſt caſt iron on which I have made experiments been taken to compute from, then inſtead of multiplying the ſtrength 25.7 by 2190, it ſhould have been multiplied by 2982, which would give 76637, for the weight it would ſupport, which divided by 9702, gives 7.89, or near 8 times the ſtrength it is loaded with.

The proprietors, on the receipt of this beam thought it too weak, and would not keep it, except the iron-maſter would warrant it for ſome years, to which he had no objections. Its weight was 9 cwt. 2 qrs. and there is no doubt but one of half the ſtrength would be more than ſufficient to work the ſame cylinder. In problem 15 we had one 3 times ſtronger than this for the work it had to do, or the reſiſtance it had to overcome. But if people chuſe to be at the expence, there is no objection, only it requires more power to move a heavy than a light beam.

On cast iron shafts or axles for mills.

PROBLEM XVII.

A water wheel is to be erected, 15 feet in diameter, 15 feet broad, and from an estimate, when loaded with water its weight will be 20 tons, what must be the side of a square shaft 16 feet long, to support that weight.

Solution.

1. Suppose the whole weight to rest upon the centre of the axles: and let z = a side of the axle sought.

Then as $1 : 2982 :: \frac{z^3}{16} : 20$ tons = 44800 pounds, from which z is found equal to 6.2176 inches, a side of the square, or a side of the axle.

2. As the greatest part or rather the whole weight of a well made wheel, will rest within 1 foot of each end of the shaft, and the strength at that distance from the end will be above 120, and the weight which it would support there, would be above 159 tons, and as much at the other end would be 319 tons.

3. Let it be required to make an axle to bear 4 times the estimated weight of the wheel at the centre, that is to bear 80 tons; then as $1 : 2982 ::$ $\frac{z^3}{16}$

$\frac{z^3}{16}$: 178920, from which $z^3 = 960$ and z, the ſide of the ſquare ſhaft is 9.826 inches.

4. The true ſtrength of an axle for this wheel, will be found by conſidering 10 tons as ſupported at the diſtance of 1 foot from the end; and the rule is, as $1 : 2982 :: \frac{z^3}{2} : 22400$ pounds, or 10 tons. $2982z^3 = 44800$, and $z^3 = 15.023$, the cube root of which is 2.4674 inches $= z =$ a ſide of the axle. As this would only juſt ſupport the weight, it would certainly be imprudent to make uſe of it, but at the ſame time proves that one a little ſtronger would be perfectly ſafe. Suppoſe it ſhould be made 4 times as ſtrong, then would $2982z^3 = 179200$, and $z = 3.9168$ inches. This is not quite 4 inches a ſide, but will bear 40 tons at a foot from the ends, and 5 tons in the centre. The ſame length, and 5 inches ſquare, will bear 10 tons at the middle, if 6 inches ſquare 13.5 tons, if 7 inches ſquare 28.5 tons, if 8 inches ſquare, upwards of 41 tons.

If the wheel is narrow and the ſhaft pretty long, ſo that the weight of the wheel may be conſidered as reſting on, or near the middle of the ſhaft, in this caſe we may compute the dimenſions of a ſhaft of equal thickneſs, and ſtrong enough to

 bear

bear 5 times the computed weight of the wheel, and afterwards diminiſh the ends from the middle, as in the engine beams, only theſe may be diminiſhed on every ſide, and the ſtrength will be as the cube of a ſide compared with its diſtance from the end, and the form will be a cubic parabola, the vertex at the end.

PROBLEM XVIII.

Let it be required to make a ſhaft 12 feet long, to ſupport a wheel of 7 tons weight.

Solution.

The axle is required to bear 5 times the weight of the wheel, or 35 tons. Then as 1 : 2982 :: $\frac{z^3}{12}$: 35 tons, or 78400 pounds, from which we find $z^3 = 315.492$; the cube root of which, is 6.8076 inches = a ſide of the ſquare axle in the middle. To find a ſide at any other diſtance from the end, ſuppoſe 4 feet, we ſay, as 6 feet is to 315.492, the cube of a ſide at the middle, ſo is 4 feet to 210.328, the cube of a ſide at 2 feet from the middle, or 4 from the end, and is equal to 5.947 inches; again, to find a ſide at 2 feet from the end, as 6 : 315.492 :: 2 : 105.164, the cube of the ſide, and its cube root 4.72 is the ſide at 2 feet from the end. Alſo for 1 foot from the end,

as

as 6 : 315.492 :: 1

1

$$6\,|\,\overline{315.492}$$

to 52.582 the cube of the ſide, and the ſide is found equal to 3.7463 inches, at one foot from the end.

The following rule may be of uſe to thoſe who cannot read algebraic theorems.

RULE.

Multiply the weight of the wheel in pounds, by the length of the axles in feet; divide the product by 2982, *and you have the cube of a ſide.*

N. B. If you wiſh to have the axle to bear 4, 5, or 6, &c. times the weight of the wheel, then you muſt multiply the eſtimated weight by 4, 5, 6, &c. and take that product for the weight of the wheel For an example, let the above problem be ſolved by it, where the axle is required to ſupport 78400 pounds, which multiplied by the length of the ſhaft 12, gives 940800, which divided by 2982, gives 315.492 for the cube of a ſide.

PROBLEM XIX.

The eſtimated weight of a water wheel is 2 tons 6 cwt. to be ſupported by a ſhaft of 9 feet long, and

and at the diſtance of 3 feet from the end, requi red the thickneſs of the ſhaft? Here it may be obſerved, that the ſtreſs at the middle, compared with that produced by the ſame weight ſuſpended from any different part of the ſame axle, is as follows; in this caſe, the ſtreſs at the middle would be as 4.5 ſquared equal to 20.25, but at 3 feet from one end, as $3 \times 6 = 18$. The weight of the wheel multiplied by 5, gives 11 tons 10 cwt. for the ſuppoſed ſtreſs at the middle, but as 20.25 : 18 :: 11 tons 10 cwt. : 10 tons $4\frac{1}{2}$ cwt. = 22897 pounds, which multiplied by 9, and divided by 2982, gives 69.105 for the cube of a ſide; and the ſide is 4.103 inches.

On the ſtrength of beams or poles of wood or iron, when uſed in the form of triangles, to ſupport weights, load waggons, raiſe ſtones upon buildings, &c.

Fig. 7, plate 3.

Let BSC be the triangle, then as the ſine of the angle which BS makes with the horizon is to the radius, ſo is the whole weight, ſuſpended from the top S, to the preſſure againſt SB and SC, for if they ſtand in the ſame poſition they will ſupport equal parts of the weight.

Thoſe who have not tables of natural ſines by them, may proportion by ſaying, as SD the altitude

tude of the pin, which supports the weight, is to the length of a pole or leg of the triangle, so is the weight suspended to double the pressure against one pole, that is when the triangle consists of two poles. If there be three poles, as for loading carriages, &c. we may say, as the perpendicular altitude of the pin is to the length of one pole, so is one third of the weight to the pressure against one pole. According to Messrs. Reynolds' experiments, (see page 89) we easily infer, that a bar of cast iron 1 inch square, and 1 foot long, will bear a pressure against the ends of about 15 tons, and it appears from other experiments, that deal, alder, and other soft wood of the same dimensions, will bear about 2.3 tons; but suppose we call it only 2 tons.

From which we get the following proportions.

For iron. As 1 : 15 :: the expression for its strength, to the weight which it will bear.

For wood. As 1 : 2 :: the expression for its strength to its load.

Examples.

PROBLEM XX.

Given, two pieces of cast iron, 2 inches square, and 16 feet long, making an angle with the horizon of 60° each, what weight will they support.

Solution.

Solution.

The expreſſion for the ſtrength is the cube of a ſide in inches, divided by the length in feet, viz. $\frac{8}{16} = \frac{1}{2}$, and the natural ſine of 60°, when the radius is 1, is .8660254, for the weight which the bar can bear againſt its end, ſay, as 1 : 15 :: $\frac{1}{2}$: $7\frac{1}{2}$ tons. next, as the length of the bar is to its perpendicular altitude, viz. as 1 : 8660254 :: 7.5 : 6.49519, which doubled, is 12.99 tons, for the whole load.

PROBLEM XXI.

Suppoſe the ſame bars make each an angle of 30° with the horizon, what weight will they bear.

Solution.

The ſtrength as before is $\frac{1}{2}$; and the preſſure which they can ſupport is $7\frac{1}{2}$ tons, the ſine of 30° is .5, and as 1 : .5 :: 7.5 : 3.75 tons for each bar, and the whole load is twice as much, or $7\frac{1}{2}$: hence, it appears that the leſs the angle which they make with the horizon, the leſs weight will they ſupport.

N. B. In round and ſquare props, poles, &c. of equal length and weight, the round will be ſtronger nearly in the ratio of 31 to 29. But if the lengths are equal, and the diameter of the round, equal to a ſide of the ſquare pole, the ſtrength of the round pole will be to that of the ſquare one as 21 to 29, nearly.

PROBLEM XXII.

Three poles 4 inches diameter and 10 feet long, make each an angle of 60° with the horizon, what load may be ſuſpended from them.

Solution.

$\frac{4 \times 4 \times 4}{10} = 6.4$ for the ſtrength, then as 1 : 2 :: 6.4 : 12.8 tons, the ſtrength of one pole. And as radius is ſine of 60° ſo is 12.8 to 11 tons, the weight which will produce a preſſure of 12.8 againſt one pole, the whole three will therefore ſupport 33 tons.

PROBLEM XXIII.

Required a triangle with 2 legs 20 feet high, and making an angle of 70° with the horizon, what muſt be their diameter to ſupport 3 tons, that is each $1\frac{1}{2}$.

Solution.

As the ſine of 70° .9396926 : radius 1 :: $1\frac{1}{2}$ half the weight : 1.5963, the preſſure upon one pole. And $\frac{x^3}{20} = 1.5963$, or $x^3 = 1.5963 \times 20 = 31.926$; $x = \sqrt[3]{31.926} = 3.17$ inches, the diameter of a pole.

PROB.

PROBLEM XXIV.

Required, a triangle with 3 legs, each making an angle of 60° with the horizon, and 12 feet long, to ſupport ſcales for weighing of 4 tons, viz. the whole load will be 8 tons.

Solution.

Firſt, as the ſine of 60° .866 &c. : radius 1 :: 2.666, one third of the weight, to 3.077 tons, the preſſure on one pole. Then $\frac{x^3}{12} = 3.077$; and $x^3 = 36.924$; hence, $x = \sqrt[3]{36.924} = 3.298$ inches, the diameter of a pole.

N. B. In the above problems, the legs will do no more than bear the weight, but if we want a triangle to ſupport 8 tons, we ought by all means to make it ſtrong enough to bear 3 times as much, and then we ſhould ſay, as .866 &c. : 1 :: 8 : 9.23. tons, and x^3 will equal $9.23 \times 12 = 110.76$; and $x = \sqrt[3]{110.76} = 4.8016$ inches.

If wood, metal, &c. intended for bows, ſprings, &c. be formed by the above proceſs, they will be equally ſtrong from end to end, but if one ſide is to be flat and equally broad through the whole length; then the other ſide is formed into a parabola, by the rule given for engine beams.

When ſhafts are placed in an upright poſition, they

they are only in danger of being broken by the twiſt, between the wheel which drives them and the reſiſtance they have to overcome. A caſt iron bar 1 inch ſquare, fixed at one end, and 631 pounds ſuſpended by a wheel of 2 feet diameter, fixed on the other end, will break by the twiſt. The ſtrength of ſquare bars in reſiſting the twiſt, is as the cube of a ſide. In round bars or ſpindles, as the cube of the diameter. Hence, if a bar of 1 inch ſquare, requires 631 pounds to break it, one of 2 inches ſquare, will require 8 times as much; and one of 3 inches, 27 times, or 17037 pounds.

I have made experiments on ſome bars of the ſame ſize, and the power or force applied in the ſame manner, which have required 1008 *pounds to break them by twiſting, but have mentioned the worſt I have tried, that I might not deceive.*

It has been obſerved, that there is no weight or preſſure to break theſe upright ſpindles beſides the twiſt, yet it may be noticed, that weight on the top, the ſhaking of the machinery, &c. tends to make them vibrate, on which account it may be well to make them ſomething thicker at the middle than at the ends.

PROBLEM XXV.

If a ſhaft 3 inches ſquare, will juſt bear the force

 acting

acting upon it, how much must a side of the square be, when it is strong enough to bear 5 times as much.

Solution.

The strength is here represented by the cube of 3 = 27, which multiplied by 5, gives 135, the cube root of which, is 5.13 inches for the answer.

PROBLEM XXVI.

Fig. 6, plate 3.

A weight W is suspended from E, the arm of a crane ABCDE, what is the pressure against the end of the spur DB?

Solution.

$\frac{CE \times W}{CD}$ = the pressure at D, in the direction DG; but the pressure at D, in the direction DB, will be as DG to DB; that is, as DG : DB :: $\frac{CE \times W}{CD} : \frac{DB \times EC \times W}{DG \times CD}$. Also the pressure against the upright post CA at B, in the direction GB, will be as BC : CD, viz. as BC : CD :: $\frac{CE \times W}{CD} : \frac{CD \times CE \times W}{BC \times CD} = \frac{EC \times W}{BC}$.

Example I.

Given, EC = 16 feet, BC or DG = 7 feet, DC = 7

$= 7$ feet, $W = 3$ tons, $DB = 9.9$; then from the above, we have $\frac{DB \times EC \times W}{DG \times DC} = \frac{9.9 \times 16 \times 3}{7 \times 7} = \frac{475.2}{49}$ 9.6979 tons, for the preſſure againſt the ſpur DB.

For the perpendicular preſſure againſt the upright axle AC, we have $\frac{EC \times W}{BC} = \frac{16 \times 3}{7} = 6.8571$ tons, the force tending to break the ſaid piece at B.

Example II.

Given $EC = 12$
$BC = 6$
$DC = 6.7$
$DB = 9$
$W = 4$

Required the preſſure on the ſpur, and the horizontal preſſure againſt the upright.

1. $\frac{DB \times EC \times W}{DG \times DC} = \frac{9 \times 12 \times 4}{6 \times 6.7} = 10.74$, the preſſure againſt the end of the ſpur. The preſſure againſt the poſt, is $\frac{EC \times W}{BC} = \frac{12 \times 4}{6} = 8$. In this example, let AC and CE be oak beams, each 10 inches ſquare, and the ſpur DB, be 6 inches ſquare. The ſtrength of EC is $\frac{1000}{10.6}$, or $94\frac{1}{3}$; which, multiplied by 660, gives 31132 pounds, which, ſuſpended at E, would break the beam CE at D. The

 length

length of the upright AC, is 12 feet, and has its ſtrength expreſſed by $\frac{1000}{12}$, which multiplied by 660, produces 55000 pounds, the weight which would break it at B. But $\frac{31132 \times 12}{6} = 62264$, the preſſure at B, which is 7264 pounds more than the beam AC can ſupport. The ſtrength of the ſpur BD is $\frac{6 \times 6 \times 6}{9} = 24$, which multiplied by 2, gives 48 tons for the ſtrength, or 107520 pounds. But $\frac{DB \times EC \times W}{DG \times DC} = \frac{9 \times 12 \times 31132}{6 \times 6.7} = \frac{3362256}{40.2} =$ 83638 pounds, which is 23882 pounds leſs than the force requiſite to break the ſpur. From the above, it appears that the upright AC is the weakeſt part; but from the principles already explained, the ingenious mechanic will eaſily proportion the parts ſo as to be equally ſtrong. I will add one example. In the above crane the horizontal beam bears 31132 pounds, the length of the ſpur and upright being given, what muſt be their dimentions, that is, how much ſquare, to be equally ſtrong as the above horizontal beam?

Firſt for the ſpur. Let $z =$ a ſide of the ſquare, then $\frac{z}{9} \times 4480 = 31132$ pounds, or $z^3 = \frac{9 \times 31132}{4480}$ $= 62.542$; and $z = \sqrt[3]{62.542} = 3.9694$ inches.

Second. The ſtrength of the upright is expreſſed

fed by $\frac{z^3 \times 660}{12}$, which muft be equal to 62264, hence $z^3 = \frac{62264 \times 12}{660} = 1132.006$, its cube root is $10.422 = z$, the fide of the fquare poft.

Let it be required to make a crane of caft iron to bear 4 cwt. but that it may be perfectly fafe, let it be calculated for 10 cwt. and let AC = CE = 3 feet, alfo BC = CD = $1\frac{1}{2}$ foot.

Solution.

Let the thicknefs of the iron be half an inch, and put z = depth of CE. Then as $1 : 2190 :: \frac{z^2 \times \frac{1}{2}}{3} : 1120$, (fee page 94) from which we find $z^2 = \frac{1120}{365} = 3.0685$; the fquare root of which, is the depth = 1.75 inches. The preffure upon the fpur at D, in the direction DG = 1120 pounds; the length of the fpur is 2.12 feet, and as DG (15) : DB (2.12) :: 1120 : 1583, for the preffure in the direction DB. As a bar 1 inch fquare, and 1 foot long, will bear 15 tons, or 33600 pounds, we fay as $1 : 33600 :: \frac{z^2}{2.12} : 1583$, from which, we find z the fide of the prop or fpur = .46385 of an inch fquare. Next, for the upright, we have $\frac{CE \times W}{BC}$, or, $\frac{560 \times 3}{1.5} = 1120$ pounds, the preffure

againft

againſt B, then as 1 : 2190 :: $\frac{x^2}{3 \times 2}$ the ſquare of the breadth to 1120 pounds, the ſame as CE, as they are of the ſame length, and the breadth will be the ſame, that is 1.75 inches.

THE

THE FOLLOWING

PROBLEMS

MAY BE OF MUCH USE TO THE INGENIOUS MECHANIC.

PROBLEM I.

IN a rod equally thick, of a given length, and loaded with two weights; to find the centre of gravity.

Put $2a =$ its whole length $= 24$, fig. 8, plate 3
$q =$ DQ $= 8$
$n =$ AM $= 4$
the weight of D $= 40$
the weight of A $= 40$
$m = 1$, the weight of an inch in length of the rod,
$x =$ GM, the diſtance of the centre of gravity from the end M.

Then $\overline{x-n} \times A + \frac{mx^2}{2} =$ momentum of the end GM, and ball A, and $\overline{2a-x-q} \times D + \overline{2a-x}$ $\times$

$\times \overline{2a - x} \times \frac{m}{2}$ expreſſes the momentum of the end GQ; from wh ch, we get $Ax + Dx + 2max = 2aD + 2ma^2 + An - Dq$. From which equation, we find $x = \frac{2aD + 2ma^2 + An - Dq}{A + D + 2ma} = \frac{1088}{104} = 10.4615 = MG$, which taken from 12, leaves $1.5385 = CG$.

Example II.

Let the weight of $D = 60$, $A = 20$, the reſt as before: then will $x = \frac{1440 + 288 + 80 - 480}{104} = 12.7$ in which caſe, G is above 12 inches from M.

PROBLEM II.

To find the centre of gyration in the ſame rod.

$2a = QM = 24$
$q = CD = 4$
$b = CA = 8$
$D = 50$
$A = 40$; $y =$ the diſtance of the centre of gyration from C.

Then will $\frac{2ma^3}{3} + Dq^2 + Ab^2$, be the force of the whole revolving round C; and is equal to the weight of the whole, multiplied by y^2; that is equal to $Ay^2 + Dy^2 + 2may^2$; and $y =$

$$\sqrt{\frac{\frac{2ma}{3} - \text{D}q^2 + \text{A}b^2}{\text{A} + \text{D} + 2ma}} = \sqrt{\frac{1152 + 800 + 2560}{40 + 50 + 24}} = \sqrt{\frac{4512}{114}}$$
$= 6.29$.

PROBLEM III.

To find the centre of percuſſion, or oſcillation in the ſame rod.

Put $b =$ diſtance of A from C the center of ſuſpenſion $= 8$, $a = 12$, $q = 4 =$ CD, $\text{A} = 10$, $\text{D} = 10$, $x =$ diſtance of the centre of percuſſion from the centre of ſuſpenſion. Then as $\frac{ma^2}{2} + \text{A}b =$ momentum of the end CM; and $\frac{ma^2}{2} + \text{D}q =$ momentum of the end CQ. The ends being of equal length, their forces will be equal, and each expreſſed by $\frac{ma^3}{3}$; and the ſum of the forces of the rod and balls, will be expreſſed by $\frac{2ma^3}{3} + \text{A}b^2 + \text{D}q^2$; which, as the balls are on different ſides of the centre of ſuſpenſion, muſt be divided by the difference of their moments $\text{A}b - \text{D}q$; for the moments of the ends of the rods $\frac{ma^2}{2}$, deſtroy each other; hence, $\frac{\frac{2ma^3}{3} + \text{A}b^2 + \text{D}q^2}{\text{A}b - \text{D}q} = x = \frac{1952}{40}$ $= 48.8$; the diſtance of the centre of percuſſion or oſcillation from the point of ſuſpenſion, ſuch a pendulum would vibrate near 60 times in 67 ſeconds, or about 53 per minute.

PRO.

PROBLEM IV.

Every thing given as in the laſt problem, except the diſtance of the ball A $= x$, which it is required to find, ſo that the vibrations may be performed in the ſhorteſt time poſſible.

Solution.

Let $\frac{2ma^3}{3} + Dq^2 = b$; then $\frac{Ax+b}{Ax-Dq}$ muſt be a minimum, its fluxion is $2Ax\dot{x} \times \overline{Ax - Dq} - A\dot{x} \times \overline{b + Ax^2} = 0$; and $A^2x - 2ADqx = Ab$, from which we find $x = \sqrt{\frac{2ma^3}{3A} + \frac{Dq}{A} + \frac{D^2q^2}{A^2}} + \frac{Dq}{A} =$ 12.39. When A $=$ D, as in this problem, the theorem may be reduced to $x = \sqrt{\frac{2ma^3}{3A} + 2q^2} + q$ If the weight of the rod is neglected $x = q\sqrt{2} + q$.

PROBLEM V.

Let ACB fig. 9. plate 3, be a pendulum, vibrating about the centre C, to find CP, the diſtance of the centre of oſcillation from C, the point of ſuſpenſion.

1. To find the centre of gravity g; divide AC and CB in the middle at D and E, and draw DE, which will paſs through g, the centre of gravity. Let S $=$ ſecant of DCg, and $r =$ radius, $a =$ AC or CB; then as S $: r :: \frac{a}{2} : \frac{ra}{2S} =$ Cg, the diſtance of the centre of gravity from the point C. And $\frac{a^2}{3}$ equal

equal the ſquare of the diſtance of the centre of gyration. Now the ſquare of the diſtance of the centre of gyration, divided by the diſtance of the centre of gravity, gives the diſtance of the centre of oſcillation, viz. $\frac{a^2}{3}$ divided by $\frac{ra}{2s}$, gives $\frac{2as}{3r}$ $=$ CP.

Example.

Let $s=4$, $a=12$, then, per theorem, $\frac{2as}{3r}=\frac{96}{3}=32$, for the diſtance of the centre of oſcillation from C, in the line CP.

2. Let the diſtance of the centre of oſcillation be given $=39.2$, to find s, half the angle ACB. We have from the above $\frac{2as}{3r}=39.2$, and $s=\frac{39.2\times 3r}{2a}=\frac{117.6}{24}=4.9=$ ſecant of $78°\ 14'$: which doubled, is $156°\ 28'$, the required angle.

3. Let the angle be required when it vibrates in half a ſecond. In this caſe, $s=\frac{3r\times 9.8}{2a}=1.225$, the ſecant of $35°\ 17'$, the angle DCg, this doubled, gives $70°\ 34'=$ the angle ACB, when it beats half ſeconds.

DESCRIPTION

DESCRIPTION OF A GAUGE

FOR MEASURING IN THE MOST ACCURATE MANNER

THE RAREFACTION IN THE CYLINDER OF A

STEAM ENGINE.

Fig. 6, plate 1.

cc a ſide of the ſteam cylinder, with a hole bored to receive the ſtop cock s, between the bottom of the cylinder and the loweſt deſcent of the piſton. The ſquare hollow piece of braſs B, is made air-tight upon the end of the cock s, by means of the nut N; and into this is ſcrewed the crooked arm ww, which has a valve at v, where there is a little water. When the plate P is ſcrewed upon the piece w, it may be ſet level by turning the piece B round upon the end of the ſtop cock s. Next we we may place the glaſs tube *t*, filled with mercury, in its baſon *b*, and cover the whole with the receiver R. Then on opening the ſtop cock T, a communication will be opened between the receiver R and the ſteam cylinder, and when a rarefaction is made in the cylinder, the air will come out of the receiver R, and ſuffer the mercury to fall in the tube *t*; but when the ſteam re-enters the cylinder, it cannot raiſe the quickſilver again, being prevented by the valve V When the mercury falls no lower, it ſhews by its altitude, the denſity of the elaſtic fluid remaining in the cylinder, and by a common barometer it may eaſily be compared with the whole preſſure of the atmoſphere. For example, if the mercury in the barometer ſtands at the altitude of 30 inches, and that in the gauge, 10 inches, then is the vapour in the cylinder equal to one third part of the weight of the whole atmoſphere. Of conſequence one third part of its weight is ſupported by this internal elaſtic vapour, and it preſſes with the other two thirds upon the piſton. The following experiments were made by concating the gauge with the cylinders, as above deſcribed, the heat of the eduction water in the hot well was alſo obſerved by a thermometer, where we could come at it.

Number

Number of the experiments.	Diameter of the cylinder.	Altitude of the gauge.	Temperature of the hot well.
1	52 in.	10 in.	130
2	40	$9\frac{1}{2}$	120
3 with a fly		7	115
4	58	9	130
5	51	11	150
6	51	$9\frac{1}{2}$	140
7	48	12	180
8	50	12	183
9	21	$2\frac{1}{4}$	Boulton and Watt, Union Steam Mill, Birmingham.
11		$2\frac{1}{4}$	Boulton and Watt, at the Eagle Foundry, Birmingham.

From theſe experiments, it appears that the length of the glaſs tube for the gauge need not exceed 14 or 15 inches.

The hot water from the cylinder aſſiſts in forming an eſtimate of the rarefaction in the cylinder, for from theſe experiments we may obſerve that where the water is hotteſt the gauge ſtands higheſt, and the contrary.

N. B. If the engine works by the force of ſteam inſtead of air, and the plate P and the glaſs gauge be removed, as alſo the valve V, and another valve fixed upon the top of the tube W; and then the inſtrument (fig. 3) deſcribed page 13 and 16, &c. ſcrewed

ſcrewed upon the end of W, inſtead of the plate P, then if the ſteam is ſtronger than the preſſure of the atmoſphere, it will be diſcovered as explained in the deſcription, and the uſe of the above inſtruments, refered to above.

Obſerve, that it may be equally convenient to have the upright part of W *to ſcrew into the horizontal part, and to have the valve at the bottom inſtead of having it at* V, *in which caſe the bottom ſcrew and valve may be turned to the top, and the ſame valve anſwers both for exhauſting and condenſing.*

FINIS.

Printed in the United States
By Bookmasters